FOUR FAMOUS NUMBERS

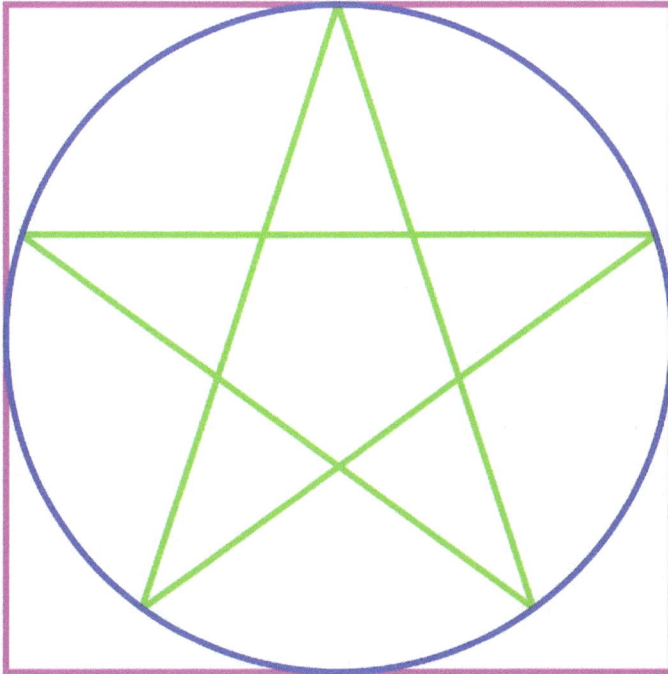

James R Warren

BLOXWICH
2023

First Published in the United Kingdom in 2023 by Midland Tutorial Productions

First Edition 1 December 2023

File Prefix Code: FFN

ISBN 978 1 915750 08 2

Midland Tutorial Productions Publishers
31 Victoria Avenue
Bloxwich
Walsall
WS3 3HS
United Kingdom

MIDLAND TUTORIAL

FOUR FAMOUS NUMBERS

First Edition

James R Warren

MIDLAND TUTORIAL PRODUCTIONS
BLOXWICH

Other Books By James R Warren

Boscawen-Ûn
Beyond Tourist Britain
Gleanings as I Pass
Exordium
Meditations
Gamma Solution
Moddeshall Hydropower
Unreasonable Mathematics
Mathematical Explorations
Researches: Volume One
Researches: Volume Two
Researches: Volume Three
Researches: Volume Four
Pi and Phi

To The Glory of
The Loving God

Who Made Our Minds Free

TABLE OF CONTENTS

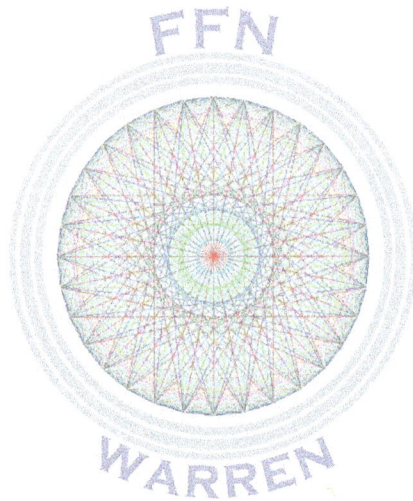

PREFACE

It is many years since I ceased teaching, or writing for money.

My work under this imprint is part of my Christian Witness. Many who read the opening passages of my previous book "Pi and Phi" will find that an odd statement, given that I there claimed that writing was often a sudden and compulsive need. But the two motives are not necessarily preclusive. A man witnesses many things he does not like, but my notional constituency is undergraduate and I write mostly of happy things.

Often I explain things in a very simplistic or even abridged manner which advanced readers undoubtedly find annoying. I suppose this is a hangover of those years of teaching adults, some of whom were graduates but the vast majority semi-skilled workers in offices or factories, who often surprised me regarding concepts they found difficult. So for my part I would rather irritate those trained to bear tedium, than lose a student. Apologies.

I have some sympathy since when I was young I was often exasperated by the many "of courses" and "clearlys" I read in scientific prose penned by my elders. I found their points far from self-evident. I have been through the script weeding these affectations, except where they genuinely contribute to understanding.

I hope at least to entertain any intelligent reader, and hopefully assist his own thought and research. By the way, in this work all references to the masculine include the feminine.

Equally I hope I fairly represent those whose honest work I have questioned, though they labor in error, as I do, as all men must. For work is faith, and faith an aspiration, and an intimation of immortality.

Most of all I hope I have done due reverence to the tiny creatures who plowed the sunless abyss of a nascent planet beneath the Gaze of God, for they cannot answer and warrant no dishonor. Some till the sea bed still. All will be Called like little Cincinnatus'es but to plow no more, much less rescue any Republic.

Most of all I hope.

The title of this book is belied by the manifest fact that I discuss many more than the four famous numbers, π, ϕ, Ψ and e, respectively the Ludolphine Constant, the Minor Ratio of Phidias, the Square Root of Two and the Napierian Base.

The Theodorus Constant, the square root of three, is also mentioned, but I was silly enough to discount it famous. You possibly call these numbers other things, but what is in a name, as a great Mercian poet once wrote?

I am not ashamed to have referenced Wikipedia or any other of the despised excellencies of our generations, though all must critically be evaluated before being broadcast abroad, as I try your work, and I hope you test mine.

I apologise to the many I have doubtless failed to thank, or merely credit.

All errors and misapprehensions are mine alone.

James R Warren
Midland Tutorial Productions
Bloxwich

19 October 2023

CHAPTER ONE
INTRODUCTION

"... no History seems to be discoverable; or only such as men give of mountain rocks and antediluvian ruins; that they have been created by unknown agencies, are in a state of gradual decay, and for the present reflect light and resist pressure; that is, are visible and tangible objects in this phantasm world, where so much older mystery is."

- *"Sartor Resartus"*,
Thomas Carlyle

Like the fitful sea mathematics is simultaneously fluid and constant, and like the sea nearly ubiquitous, with latent powers and hidden depths.

Like the fitful sea mathematics is a rolling symphony of repetitions with variations.

Like the fitful sea mathematics presents a glaucous plain or roiling cover for pelagic mysteries above a bed of alien conformation, unintelligible in part, unknowable in whole, yet complacent in its solicitous infinities, mutable and static.

Within its shifting stories dwell strange denizens Answerable only to their Creator though prey to strange predators.

Take for example the modest sea-slug, an omnibus name for several benthic phyla. Men and women barely knew him an hundred years ago, except for his edible shallow-water brethren. Often gaudily colored, or else garbed in the most delicate mauves and whites, he quietly plows the mud and slime of the sea floor in a deep and unremitted darkness where no color beguiles men, if ever it was meant to.

The mighty whale sports and gambols his season near the shining interface that coruscates mercurially miles above. At last he dies, and falls through the water column to his final bed. Here the gentle sea-slug finds him and assists his allies the crabs and shrimps kindly to dismember the whale and transition him to further mysteries.

You possibly object these are matters of faith or of factual science, meat for oceanographers and marine biologists, whilst mathematics is a literary figment, expressive of what mankind wishes to make of it.

Well, that may be so.

But think of another strange creation:-

$$T = e^{\frac{\pi\Psi}{\phi}}$$

Equation 1.1

Like the sea-slug, this confection was known only in part to the argonauts of olden days who literally or figuratively cruised the sunlit surface of things.

Within this seemingly meaningless congeries of glyphs lie intestine complexities in their own ways as involved as the anatomies of sea-slugs, or indeed greenhouse cucumbers.

Take for example the Ludolphine Constant, π. You may define it in a very simple way and leave the matter at that. You may declare, without fear of contradiction, that it is the Ratio of the Length of a Circle's Perimeter to the Length of its Diameter. You may attach an approximate number value to this thought-object, for example the value 22/7, which in turn approximates 3.14285714285714. Already we think we decern that there is detail and complexity hidden within π: It has structure as well as substance, form as well as content. We suspect that there may be even greater documentable depth. For example we can see at once that the digits after the decimal point repeat in groups of six. This is deceptive. The *actual* digits of actual π do not repeat in groups. The first twenty-one digits of true π are 3.14159265358979323846. There is no pattern, and no pattern repeats.

With apologies to the late Carl Sagan, as we enter our submarine of the imagination and dive deeper the World unpeels like a Russian doll, if you will further forgive my clumsy metaphors, to disclose an infinite regress of discovery, comprehended by the Infinite, but only observed, and that in part, by the explorer.

Helminthoida labyrinthica, Nereites irregularis and his Cousins[1.1]

On the Paleozoic sea-floor tiny creatures drew traces that have come down to us as grooves on the bedding-planes of ancient sedimentary rocks. Think of the creatures as little plowmen tilling the soil of the sea bed: Our "sea-slugs".

The term "sea-slug" is used by me in a very vague and generic manner, and covers diverse phyla of similar habitat and behavior, extinct and living, some of whom have never been seen in the body, fleshy or fossil.

The marine biologist's "sea-slug" or "sea-cucumber" is only one of many modern animals capable of drawing grooves on or slightly inside the benthic silts and muds. Ancient and modern phyla of animals capable of such behavior include Annelida (true worms); Holothuria (true sea-cucumbers); and even Arthropods (crabs, shrimps, etcetera). This list is by no means exclusive. I am easily insolent enough to assume that you are a Chordate, and if so your kind has graven many traces on the ocean bed, sometimes intentionally.

The Latin names enjoyed by these long-dead "sea-slugs" reflect the confusion of men. When a mollusk or arthropod authors these grooves his exoskeleton very occasionally persists in a fossil form at the terminus of a trace. But it is almost invariable that the creature is absent, and thus presumably soft-bodied.

The word _Helminthoida_ is highly-prejudicial and infers that the groove-cutters were worm-like. Though the labyrinth is another Classical metaphor it is only descriptive of the trace-fossil as it has come down to our time. Only the behavior of the animal has survived him in the form of works. On the other hand Nereites, or "Nereid Stone" is tantamount to a confession that we do not know who or what made these tracks. A Nereid was a Greek sea-goddess, a daughter of Doris.

Figure 1.1 is a photograph[1.2] taken from the Wikipedia article "Nereites" and displays a congeries of Nereites irregularis on a bedding-plane of ancient rock. (I have lowered the gamma of the photograph so as to highlight the trace-fossil pattern).

From a mathematical point of view it is rational to assume, in the absence of compelling science to the contrary, that the organism sought to optimise its feeding strategy by maximizing the ground plowed in a given area of sea-bed. Note the ontological minefield laid for us by concepts such as "rational", "optimise", "organism" and even "science".

This is analogous to mariners or aviators or other systematic searchers scanning ground in a Boustrophedon pattern of parallel rows, or alternatively an arithmetic spiral pattern devolving from an assumed center of last-contact.

Most people who have thought about the matter have long concluded that the boustrophedon is the most efficient way of covering ground. The boustrophedon is the immemorial pattern of the plowman who tills the earth with team or tractor, and who turns his tackle at a telson to make another parallel groove in the opposite direction. I confirmed that the boustrophedon is better than the arithmetic spiral for covering the ground in my 2022 publication "Researches: Volume Three"[1.3] and that it is also superior to Hilbert-type fractal patterns, though not necessarily to all fractals.

Notwithstanding that, the organism who generated Nereites seemed himself uncertain about which was the most efficient way of plowing his patch, hedging between the boustrophedon and the spiral, or at least sectors of spirals.

So the organism of Nereites was *in two minds*, or if you prefer, confused. Confusion is the privilege of organisms who have a brain. The Holothurian has no brain: The Annelid does.

Does this prove that we can learn from slugs and worms, even those who bequeath us a legacy from the primordial abyss?

At the expense of a degree of caricature I have schematised typical Nereites in Figure 1.2 in order to clarify the character of a hybrid spiral-boustrophedon pattern.

Figure 1.1
Nereites irregularis
On the Bedding-Plane of an unidentified
Paleozoic Argillaceous Sandstone

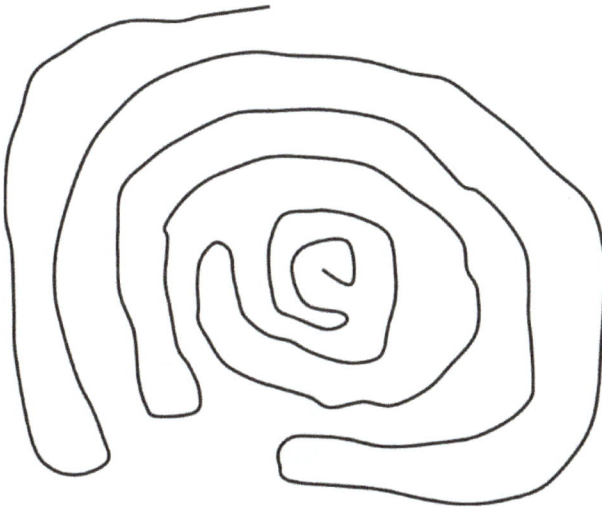

Figure 1.2
Schematic Diagram of a typical Nereites irregularis
Trace Fossil of Paleozoic facies

Gendering

Since earliest infancy I have embraced the pious prejudice that all living creatures should be credited with sex, and that in the absence of definite information thereto the masculine should be assumed, at least for the purposes of grammatical gender.

In my nonage my Mother would invariably call an animal "he" with the curious exception of a pussy cat, who was invariably feminine. In these earliest years I seldom saw my Father, who was usually at sea, but when I did see him he would speak of little else but ships who were invariably "she". I write of the middle years of the last century. Britain was a very different place.

Gender is not a natural feature of English nouns, and Englishmen tend readily to confuse sex and gender, and even gender and sexuality.

Sex is an attribute geared to reproduction of the phenotype and is therefore propagative. Sex is instrumental. Things that never die have no use for reproduction of their type. God never dies and therefore it is not logical to impute Him sex, though we invariably style Him Him.

Without comparative intent, we suffer major ontological problems when we discuss trace fossils, or even fresh tracks.

A trace fossil or indeed a fresh track is not an animal; it may not even be the vestige of an animal. It is what it says, a trace or grove or linear disfigurement of the surface of a rock, or a modern lake or sea bed. It would be inappropriate to gender this, and absurd to ascribe it sex.

Benthonic invertebrates are of ambiguous sex, at least to human eyes. The polychaete annelid typical of marine sediments is asexual whilst oligochaetes reproduce asexually in the summer and sexually in the autumn, though this appertains to terrestrial species, for how could a creature distinguish season in the monotone fastnesses of the abyss? Mollusks may be of either sex or hermaphrodite and arthropods are definitely sexual.

Sometimes the dead agent of incision is found at the end of a trace fossil. It is invariably accused of drawing the glyph and if it is hard enough for its form to persist then it is of a sexual phylum.

In our modern world, I mean since the Stone Age days of dug-out canoes, mankind has busily been scribing trace fossils on the mud and sand of lake and sea beds, mostly in shallow water. Anchors have been dropped and dragged and chains laid. The tide and currents have then orbited the chains about the mooring. Trawls have been dragged and traps laid.

On an estimated three million occasions, on the British and Irish shelf alone, whole ships have made a final voyage to the benthic zone where they have graven troughs in the sediment, swilled craters in the silt, and formed dunes of current-borne sand in their lee.

I say three million, but the number is innumerable. The ocean is an abyss, dark, silent and remote, finite but immense.

Mathematics is dark and silent, but available, amenable and infinite.

One is irresistibly reminded of the distinction between the engraver and his work, between the Nuomenal and the Phenomenal, the Genotype and the Phenotype, the temporal and the indefinite, the active and the propagative and the passive and the still. Of Archimedes' lever; of Newton's equal and opposite forces.

The Paradox of Passive Agency

Many of you assert that mathematics is the Queen of Sciences, indispensable to the understanding and management of natural Sciences, and even more obviously of Applied Sciences such as engineering and medicine.

And that it is accordingly of most signal and indispensable utility.

I most assuredly agree, at least in the spirit of the claim, if reluctantly not in the letter.

You see, theoretically speaking, mathematics is a textual language, the distant cousin of Cornish, English and Mandarin. Many of the symbols and structures of modern mathematics were developed by European savants within the last five hundred years to solve very specific, and often very mundane, commercial or navigational problems, such as gambling issues and insurance.

To an extent, Ancient mathematicians thought to relate shapes and angles rather than textual analogies. In other words their mental habit was geometric rather than algebraic, but the two styles can systematically be compared, indeed related.

Of course, we can discuss scientific topics or tropes in English or French or any language for which intelligible imagery is possible, even graphical presentations or signals. But without mathematical methods discussion is clumsy and limited in scope.

When mathematics is driven by human skill it becomes potent and this is what it means to call it a passive agent, rather like a steam locomotive that demanded the concerted skills of at least two men to make it move. Modern machines are designed upon mathematical principles, and without mathematical machines the topics we address would not be manageable within a human lifetime, given that our lives are both more frantic and more brief than either the lives of the whale or the sea-slug.

The Haunted Scaffold

As we remarked above, the Ancients, including The Greeks often thought in pictorial terms. For example, simple two-dimensional shapes such as the Circle, the Square and the Pentagram were seen to include strange properties, which in turn could be described by equally perplexing numbers.

The Pentagram is a five-pointed star-like figure, very suggestive of the Asterozoa (starfish) who neighbor the sea-slugs upon the sea-bed.

The pentagram, as a two-dimensional line construct, was known to the Ancient Sumerians and Babylonians at least as far back as 3500BC, or vaguely the Eneolithic Era as it applied in the Mesopotamian region. It is most likely that the Periclean concept of The Ratio of Phidias, Φ, was evolved from the study of the pentagram by Greek mathematicians, because the lineaments of the pentagram form a framework or scaffold "haunted" by Φ in various guises.

The (Major) Ratio of Phidias, Φ, (The Extreme and Mean Ratio of Euclid), also known as the Golden Section, or other arbitrary names, is conveniently defined by the algebraic formula:-

$$\Phi = \frac{a+b}{a} = \frac{a}{b} = \frac{1+\sqrt{5}}{2} \approx 1.61803398874989$$
Equation 1.2

Also, the Minor Ratio of Phidias, ϕ, is given by:-

$$\phi = \Phi - 1 = 0.61803398874989$$
Equation 1.3

The ratio may be constructed geometrically by a simple compass-and-straightedge partition about a square. For details, please consult a standard text such as my book "Pi and Phi"[1.4].

Respectfully let me assure you, if assurance is necessary, that there is nothing metaphysical about the pentagram or any of its parts. It is the toy of neither God nor Satan. It is only a figure that men draw in aid of their designs and deliberations. For sure, The Ratio of Phidias Φ and its conjugate φ describe all the straight lineaments of the figure, in association with other constants, but this is only the logical outcome of the laws of trigonometry that relate the ratios of line lengths to the angle between two lines. This trigonometry serves, so to say, as a bridge between the purely graphic concepts of Geometry beloved by the Ancients: And the textual Algebraic expressions familiar to Islam and the modern West. As this history implies, all these mathematical objects are inventions of mankind.

Figure 1.3 depicts three key lines within the dark green pentagram, and relates them to such constants. Other key metrics are excluded for clarity. To manage such relationships it is convenient to circumscribe the pentagram with a (larger) circle (shown in blue) and a square (depicted in purple). Also, it is helpful to define the larger circle's diameter as unity (i.e. its radius is ½) which implies that the square has unitary side length.

You can choose whatever datum length you prefer, but non-unitary diameters are merely troublesome: They provide no extra information.

The anatomy of the square and its inscribed circle and pentagram will be discussed in more detail later.

But at this juncture we may introduce the Pythagoras's Constant, Ψ, which is the Square Root of Two.

The Pythagoreans were a fraternity of mathematical magi who followed the mystic Pythagoras in the Greece of about 500BC, precursors of the philosophers of the golden age of Periclean Greece some hundred years later. The Pythagoreans believed that all numbers could be expressed in a (literally) Rational manner as the ratios between two integers, for example 22/7 or 1414/1000. One of their brethren, Hippasus of Metapontum, demonstrated that this was not the case for the square root of two. For this heresy Hippasus was jettisoned (literally). The approximate value of the square root of two is 1.4142135623731, but the digits after the last never terminate because Ψ is "Irrational".

As an intelligent reader you have probably gathered by now that the square root function is key to understanding these relations between mathematical objects and that accordingly any engine that can be devised to compute square roots rapidly and accurately is something earnestly to be sought.

The Square Root Function

Functions are a class of "black box" mathematical tools into which you can plug pre-cursor numbers or a single number, and compute an outcome.

There are an infinite number of possible functions to compute anything given any input.

Naturally, as practical men and women, we quest for efficient and effective functions that give us something industrially-useful as quickly as possible. Not all functions are equal.

The Newton-Raphson Method

The Newton-Raphson Method for the iterative calculation of a (square) root is a reasonably simple and accurate function that repeatedly processes its own output until the discrepancy between the latest root estimate and the penultimate one falls below a predetermined tolerance.

Allow that the Root for Radix radix and Argument arg is defined by:-

$$Root = \sqrt[radix]{arg}$$
Equation 1.4

For example, the fourth root of 27 is:-

$$Root = \sqrt[4]{27} = 2.279507056954780$$
Equation 1.5

then the Newton-Raphson iterative succession is defined as:-

$$x_{i+1} = x_i - \frac{x_i{}^{radix} - arg}{radix \times x_i{}^{radix-1}}$$
Equation 1.6

The phrase $x_i{}^{radix}$ - arg is the Function $f(x_i)$ whilst the denominator $radix \times x_i{}^{radix-1}$ is the First Differential of $f(x_i)$, often denoted in Newtonian script as $f'(x_i)$ or in Leibnitzian conventions dx/dy.

Therefore this form is often seen in literature:-

$$x_{i+1} = x_i - \frac{f(x_i)}{f'(x_i)}$$

Equation 1.7

The Newton-Raphson Method is very robust and very economical as long as it is applied judiciously, and in particular that the starting point x_0 is well-chosen, especially when multiple roots are present, though this last is unlikely to embarrass simple root-finding.

The Monotone Convergence Theorem

The Monotone Convergence Theorem for the Square Root was discovered in Mesopotamia around the year 3500BC: As far as Europe is concerned this is definitely in the Stone Age, but the Mesopotamian peoples of the time possibly enjoyed a primitive metallurgical culture using arsenical copper to make tools and weapons.

In formal, modern terms the Monotone Convergence Theorem for the Square Root is defined by:-

$$y_{i+1} = \frac{1}{2} \cdot \left(y_i + \frac{a}{y_i} \right)$$

Equation 1.8

where y_{i+1} is the Successor Estimate to y_i, and a is the Argument.

As you see, The MCT is even simpler than the Newton-Raphson Formula and hardly inferior, delivering thirteen-figure accuracy after five iterations. The remarkable thing is that five thousand years separate the two procedures.

The Euler Convergent

When the Euler Convergent is applied to the Taylor Series Expansion for the expression $(1+x)^{(1/\text{radix})}$ we are able to define:-

$$Sqrt_{euler} = \sum_{k=0}^{n} \frac{(2k+1)!}{2^{3k+1} \times (k!)^2}$$

Equation 1.9

Though this equation is a century later than the Newton-Raphson Method it is markedly inferior to that and the MCT both in terms of speed and output.

This shows, if any exposure were needed, that the latest methods are not always the best methods, even when world-historical geniuses are at play.

Table 1.1 lists the convergent successor values for all three of these methods demonstrating that for the two iterative methods very good estimates of Ψ are achieved within five or seven applications.

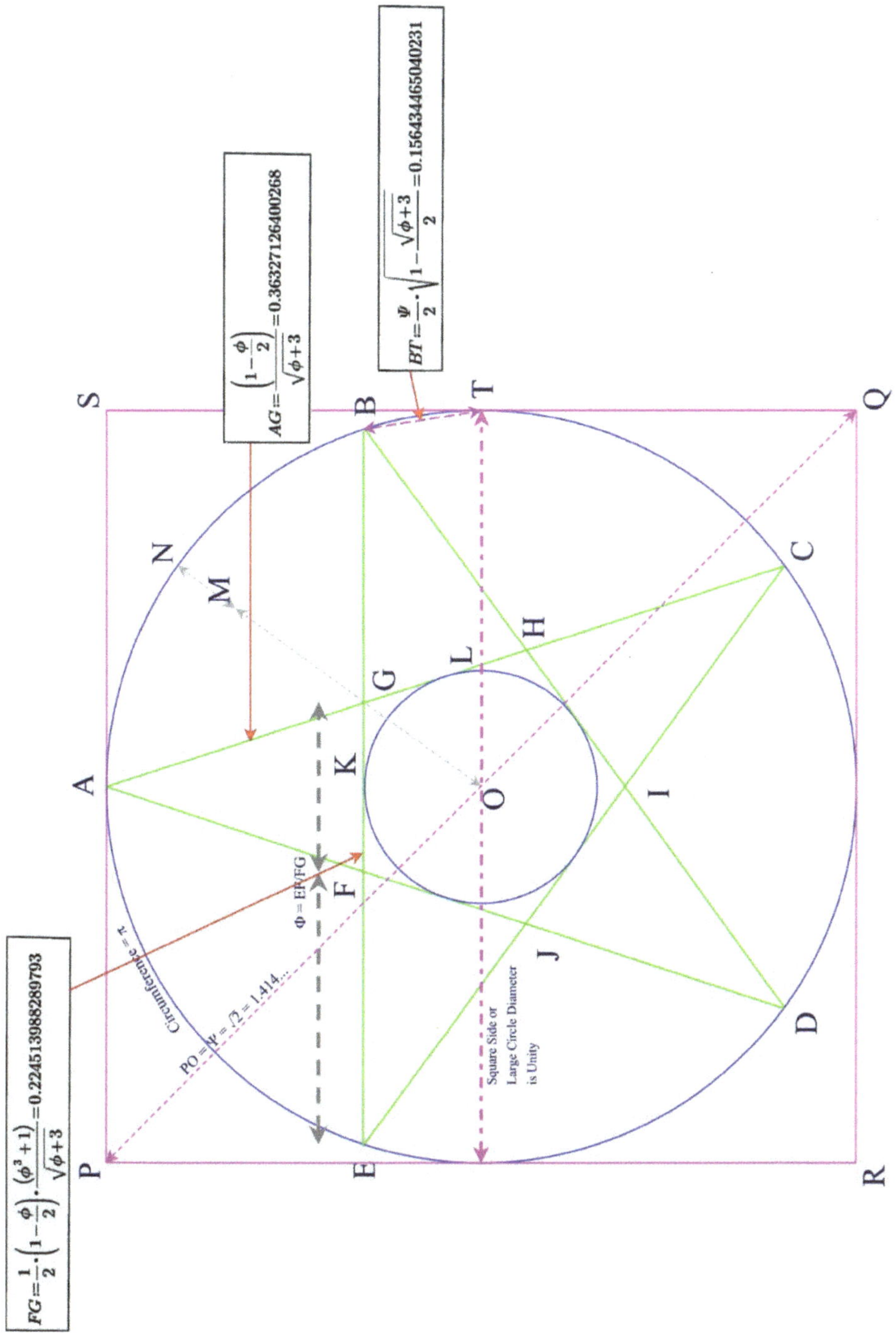

$FG := \frac{1}{2}\cdot\left(1-\frac{\phi}{2}\right)\cdot\frac{(\phi^3+1)}{\sqrt{\phi+3}} = 0.224513988289793$
$AG := \frac{\left(1-\frac{\phi}{2}\right)}{\sqrt{\phi+3}} = 0.36327126400268$
$BT := \frac{\Psi}{2}\cdot\sqrt{1-\frac{\sqrt{\phi+3}}{2}} = 0.156434465040231$
S
B
T
Q
N
M
C
H
L
G
A
K
O
I
F
J
D
P
E
R
Φ = EF/FG
Circumference = π
PO = Ψ = √2 = 1.414...
Square Side or
Large Circle Diameter
is Unity

Figure 1.3
The Minor Ratio of Phidias, ϕ,
(and the Pythagoras's Constant, Ψ)
in Parts of the Pentagram

Ψ_{fido} 1.41421356237310

x_0 1.5

Serial k	3500BC Monotone Convergence Theorem	1669AD Newton Raphson Method	1768AD Euler Convergent using Taylor Series
0	1.5000000000000000	1.5000000000000000	0.5000000000000000
1	1.41666666666666670	1.41666666666666670	0.3750000000000000
2	1.4142156862745100	1.4142156862745100	0.2343750000000000
3	1.4142135623746900	1.4142135623746900	0.1367187500000000
4	1.4142135623731000	1.4142135623731000	0.0769042968750000
5	1.4142135623731090	1.4142135623731090	0.0422936328125000
6	1.4142135623731000	1.4142135623731000	0.0229110717777344
7	1.4142135623731090	1.4142135623731090	0.0122737884521480
8	1.4142135623731090	1.4142135623731090	0.0065204501152040
9	1.4142135623731090	1.4142135623731090	0.0034413486719130
10	1.4142135623731000	1.4142135623731000	0.0018067080527540
Total			1.4122487772225610
PSD	0.0000000000000016	0.0000000000000000	0.1389312901358 82

Table 1.1
Three Methods for the Square Root of Two Compared

The First Differential

The First Differential of a function is the rate of change of the Dependent Variable ("y") with regard to the Independent Variable ("x").

Colloquially speaking, it is the slope or Grade of the line drawn by plotting y in the ordinate (vertical) direction against x in the abscissa (horizontal).

The First Differential is the most basic process in calculus. Its converse is Integration, essentially the computation of area under the functional curve.

When f(x) was y, the function of x, the habit of Newtonian mathematicians was to write the First Differential as f'(x) (f prime of x). This subtlety could confuse printers and readers. The Leibnitzian (German) convention was more graphically to write dy/dx. I.e.:-

$$First\ Differential = \frac{dy}{dx} = f'(x)$$
Equation 1.10

dy can be thought of as a vanishingly small Infinitesimal Increment along the vertical where dx is the corresponding increment along the horizontal.

The First Differential is extremely useful when you are researching number formation.

Scientists and theoreticians use it to model physical situations. For example, velocity is the rate of change of position with regard to time, say for sake of argument in kilometers per hour. By further differentiating d(position)/d(time) in an Ordinary Differential Equation physicists can define acceleration as the rate of change of the rate of change of velocity. Thus acceleration is a Second Differential, f" or d^2y/dx^2.

In empirical research it is not always obvious what constitutes an accurate estimate of line-slope at a given point, and at some positions the slope may be effectively zero (the graph is horizontal) or infinite (the graph is vertical).

The simplest function that can be differentiated or integrated is the Algebraic Polynomial, essentially a sum of weighted powers of x which represents y at given points. An Algebraic Polynomial always has as many real or complex Roots as its highest Degree (power of x). A number like π which is the root of no polynomial is described as being a Transcendental Number. Ψ is not a Transcendental Number, and neither is Φ: They are roots of different polynomials.

In formal terms:-

$$P(x) = \sum_{i=0}^{n} c_i x^i$$

Equation 1.11

where P(x) is an Algebraic Polynomial in Independent Variable x; n is the Degree of the Polynomial, i.e. the number of summative terms; and c_i is the Coefficient (weight) attaching to the term x^i. c_i is capable of being zero.

For example, if c_0 is 2.333, c_1 is 0.2, and c_2 is 14 we can expand our Polynomial as the Quadratic Equation:-

$$P(x) = 14x^2 + \frac{1}{5}x + 2.333$$

Equation 1.12

In terms of its application to the Algebraic Polynomial the First Differential may be generalised as:-

$$P'(x) = \sum_{j=1}^{n-1} j \cdot c_j \cdot x^{(j-1)}$$

Equation 1.13

For example, the degree-two Equation 1.12 has the degree-one analytical First Differential:-

$$P'(x) = \sum_{j=1}^{n-1} j \cdot c_j \cdot x^{(j-1)} = 1 \times \frac{1}{5} \times x^0 + 2 \times 14 \times x^1 = 28x + \frac{1}{5}$$

Equation 1.14

The outcome P'(x) = 28x + 0.2 defines a First-Degree Polynomial which is geometrically a straight line.

We could try to fit a straight-line to the first differentials of a series of convergent terms defining some approximation of the square root or something.

The observed result may or may not be satisfactory. A number of vagaries may supersede: Our method of measuring the local slopes may be inadequate; our computer may insinuate errors of its own; the assumption of our scientific model may be wrong, for the particular process under review, or for the family of like programs for calculating the function of interest.

We will assess a local First Differential f'5pt$_j$ in terms of its four neighboring points f$_{j-2}$, f$_{j-1}$, f$_{j+1}$, f$_{j+2}$, so that the numerically-estimated first differential locates at Point j.

The appropriate five-point equation is:-

$$f'5pt_j = \frac{f_{j-2} + 8f_{j-1} - 8f_{j+1} - f_{j+2}}{12.\Delta h}$$

Equation 1.15

where f(j) is the Value of the Original Convergent Function at Point j; and Δh is the (equal) Interval between j and j±1. In our work the Interval Δh is always unity. The Δ sign has been included to emphasise that h is an interval, but the Δ is redundant and you usually only see Δ or h alone in literature.

This Equation 1.15 was found empirically to agree exactly with analytic results.

It is also empirically clear that the First Differential of a Convergent Series can be approximated by the logarithm of a First-Degree equation fitted using linear regression or some other approximation tool. The base of the logarithm concerned is not material because it is adjustable using a scaling constant such as log$_{10}$(e) or some such.

Eventually, I chose to standardise on the Napierian Base, e, which will itself be the object of convergence studies in due course.

Accordingly our conjecture is that:-

$$\ln(|f'(x)|) \approx \pm U - Vx$$

Equation 1.16

where ln() is the Natural (Napierian) Logarithm function; |....| is the Absolute Value Operator; U is the Convergent Term Complexity; and V is the Logarithmatised Rate of Convergence. If the RHS negative sign is replaced by a plus in the context of any solution series then the series diverges and is useless for determining the value of a number.

The Absolute Value operation is needed because an f'
value often proves negative, especially for osscilative series, and there
is no logarithm of a negative number. Rendering all f' as positive does
not vitiate the integrity of approximation of Equation 1.16

As a rule we seek a procedure or algorithm where U is
absolutely as small as possible and V is as large as possible, subject to
computational constraints.

By the inversion law of logarithms it follows from
Equation 1.16 that:-

$$|f'(x)| \approx e^{(\pm U - Vx)}$$
Equation 1.17

<u>Empirical Illustrations</u>

At this stage students may find brief tabulated
illustrations helpful.

We will look at the behavior of three historically
important convergent series for the square-root of two, Ψ:-

(a) The Monotone Convergence Theorem
     This is the Bronze Age "Babylonian" Method
     defined by Equation 1.8
(b) The Newton-Raphson Iteration
     The seventeenth-century algorithm defined by
     Equation 1.7
(c) The Euler Method
     This is the application of the Euler Convergent
     to enhancement of the Taylor Series expansion
     of $(1+x)^{\frac{1}{2}}$. This was developed from century-old
     English calculus principles by the eighteenth-
     century St Petersburg-based Swiss genius,
     Leonhard Euler. In a modern idiom it is defined
     by Equation 1.9

Table 1.2 is essentially a version of Table 1.1 represented for ease of comparisons. You see immediately that the
Babylonian and Newton-Raphson iterates are identical: Indeed the
former is merely a special case of the latter for the radix two. The Euler
Process is different in conception since like the Taylor Series its basis
approaches the root by *summation* of the convergent series terms.

Table 1.3 exhibits the five-point first differentials of the
Table 1.2 entries. Note that the first two and last two entries of each
series are undefined according to Equation 1.15. In general f' may be

negative or positive: Accordingly, we have to use the absolute values of these numbers. It is a fortuitous convenience that in this example all the f' values are naturally positive.

In a convergence successive f and f' values diminish.

		Approximate Date of Invention	3500BC	1669AD	1768AD
Ψ_{fido}	1.414213562373100				
x_0	1.5				
		Serial k	Monotone Convergence Theorem	Newton Raphson Method	Euler Convergent using Taylor Series
		0	1.500000000000000	1.500000000000000	0.500000000000000
		1	1.416666666666670	1.416666666666670	0.375000000000000
		2	1.414215686274510	1.414215686274510	0.234375000000000
		3	1.414213562374690	1.414213562374690	0.136718750000000
		4	1.414213562373090	1.414213562373100	0.076904296875000
		5	1.414213562373090	1.414213562373090	0.042297363281250
		6	1.414213562373090	1.414213562373100	0.022911071777344
		7	1.414213562373090	1.414213562373090	0.012273788452148
		8	1.414213562373090	1.414213562373100	0.006520450115204
		9	1.414213562373090	1.414213562373090	0.003441348671913
		10	1.414213562373090	1.414213562373100	0.001806708052754
					1.412248777225610 Total
			0.000000000000016	0.000000000000000	0.138931290135882 PSD

Table 1.2
The Convergence of Historic $\Psi=2^{\frac{1}{2}}$ Iterations
and the Euler Summation

Valid logarithms of the first differentials according to Equation 1.16 are always negative.

Using EXCEL® I fitted first-order algebraic polynomials ("straight lines") to each series using linear regression. Remember that statistical regression fittings are always *only models* or analogies and exact co-incidences cannot be expected.

Table 1.3 confirms the identity of the Babylonian and Newton-Raphson methods in the square-root context. The logarithmic values are absolutely much smaller for the Euler Method showing that the method is far inferior to the Babylonian and NR Methods: As with jests and men, the old ones are sometimes the best ones, even as we said earlier when it is the case that the jokers are world-historical geniuses. One is compellingly reminded of the over-previous General Relativity paper that a young Einstein posted on the window of

Selfridge's store, only to rip it down days later when he realised his error.

Ψ_{fido}	1.414213562373100				
x_0	1.5				

Approximate Date of Invention		3500BC	1669AD	1768AD
	Serial k	Monotone Convergence Theorem	Newton Raphson Method	Euler Convergent using Taylor Series
	0			
	1			
	2	0.008784272663560	0.008784272663560	0.194112141927083
	3	0.000205841292074	0.000205841292074	0.132705688476562
	4	0.000000176992848	0.000000176992848	0.080569585164388
	5	0.000000000000133	0.000000000000133	0.046365896860759
	6	0.000000000000000	0.000000000000000	0.025881037116051
	7	0.000000000000000	0.000000000000000	0.014165082325538
	8	0.000000000000000	0.000000000000000	0.007646990163873
	9			
	10			

Table 1.3
The Convergence of Historic $\Psi=2^{\frac{1}{2}}$ Iterations
and the Euler Summation:
The Five-Point First Differentials f'

Figure 1.4 is a plot of the Natural Logarithm of the Absolute First Differential, $\ln(|f'|)$ versus the Serial Number, k, of the Convergent in x for the weak Euler-Taylor system and the stronger Newton-Raphson or its equivalent Monotone method. In the latter case, a linear regression is a poor fit to the plunging conformation, and a quadratic regression would be hardly better, but at this stage all we require is an *approximation* of Equation 1.16:-

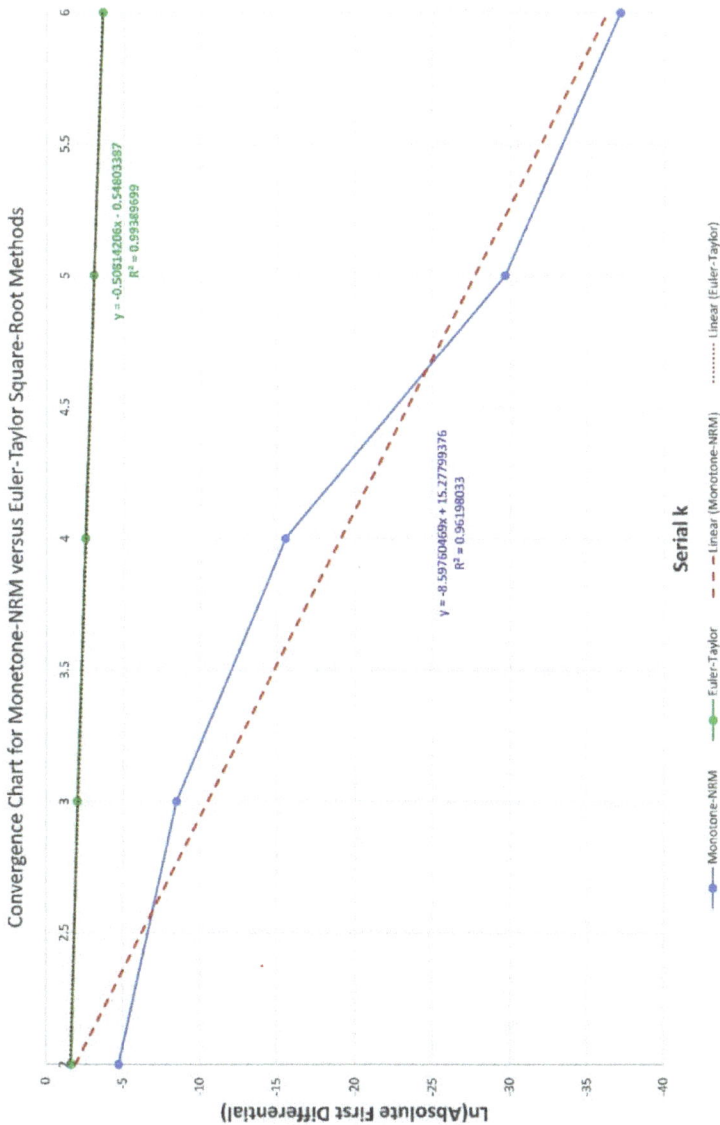

Figure 1.4
The Convergences of the Monotone-NewtonRaphson
and the Euler-Taylor Series
For Square Root Approximation
Together with their respective Linear Regressions on
Number of Convergents, k, versus Ln(|f'|)

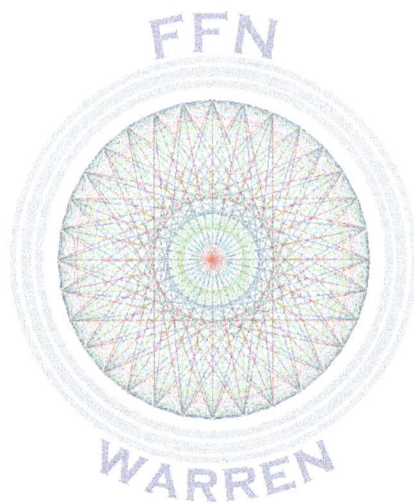

CHAPTER TWO
NUMBERS IN A PENTAGRAM
AND ITS ASSOCIATED SHAPES

We may now develop some of our arguments about numbers "hidden" in the geometry: Arguments in both the colloquial and the mathematical senses.

As remarked in Chapter One, the spatial geometry of lines and angles can be expressed as algebraic numbers and equations mediated by trigonometry. The trig is itself expressible as algebraic series, etcetera, and in particular as often simple surd arrangements of integers and numerical roots.

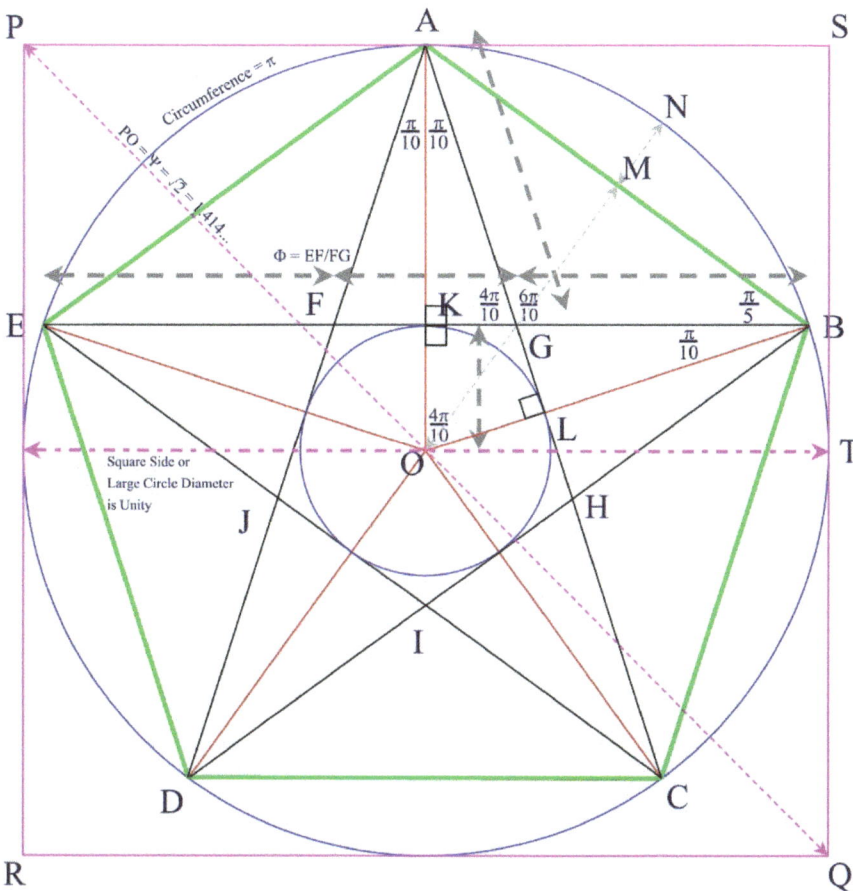

Figure 2.1
The Pentagram and its Circumscribed Circle,
with the
Circle Circumscribed by a Square

Some Key Angles

To illustrate, KBO = π/10 is an angle one tenth of the sweep of a circle, i.e. it is 36°, and we may re-express it as:-

$$\sin(KBO) = \sin\left(\frac{\pi}{10}\right) = \frac{\sqrt{5}-1}{4} = \frac{\phi}{2} = \cos\left(\frac{4\pi}{10}\right)$$

Equation 2.1

whilst:-

$$\cos(KBO) = \cos\left(\frac{\pi}{10}\right) = \frac{\sqrt{10+2\sqrt{5}}}{4}$$

Equation 2.2

As per our usual convention, φ is the Minor Ratio of Phidias, whose numerical value is about 0.618033988749895

The numerical value of sin(π/10) is 0.309016994374947 and of cos(π/10) is 0.951056516295154

The value of obtuse angle ABC, one of the internal angles of the circumscribed regular pentagon, is yielded by:-

$$ABC = 2\left(\frac{\pi}{5} + \frac{\pi}{10}\right) = 2\left(\frac{3}{10}\pi\right) = \frac{3}{5}\pi = 1.88495559215388$$

Equation 2.3

Lengths

Because the Circle Diameter, 2.AO, is unity by definition, and is the same as the Square Side Length, PS, it is manifest that the Circle Radius, AO, is one half, i.e. 0.5 = ½

Accordingly we may specify the Minor Inscribed Circle Radius, r = KO as:-

$$KO = r.\sin(KBO) = \frac{1}{2}.\sin(KBO) = \frac{\phi}{4}$$

Equation 2.4

whilst:-

$$KB = r.\cos(KBO) = \frac{1}{2}.\cos(KBO) = \frac{\sqrt{10+2\sqrt{5}}}{8} = \frac{\sqrt{\phi+3}}{4}$$

Equation 2.5

Also, by the Theorem of Pythagoras:-

$$\Psi = PQ = 2\sqrt{AO^2 + AO^2} = 2\sqrt{\left(\frac{1}{2}\right)^2 + \left(\frac{1}{2}\right)^2} = 2\sqrt{\frac{1}{2}}$$

Equation 2.6

Ψ, the Square Root of Two, is another of our Four Famous Numbers, joining π and ϕ in our haunted scaffold. It is likely to be much more difficult to detect the rôle of e, the Napierian Base, in this framework.

Furthermore:-

$$AK = r - KO = \frac{1}{2} - \frac{\phi}{4} = \frac{1}{2}\left(1 - \frac{\phi}{2}\right)$$

Equation 2.7

Now by the Sine Rule:-

$$AG = \frac{AK}{\frac{\sqrt{10 + 2\sqrt{5}}}{4}} = \frac{4.AK}{\sqrt{10 + 2\sqrt{5}}}$$

Equation 2.8

since:-

$$\sqrt{10 + 2\sqrt{5}} = \sqrt{10 + 2(2\phi + 1)} = 2\sqrt{\phi + 3}$$

Equation 2.9

it follows that:-

$$AG = \frac{4.\frac{1}{2}.\left(1 - \frac{\phi}{2}\right)}{2\sqrt{\phi + 3}} = \frac{\left(1 - \frac{\phi}{2}\right)}{\sqrt{\phi + 3}}$$

Equation 2.10

The internal line element KG is then expressible as:-

$$KG = AG \times \frac{\sqrt{5} - 1}{4} = \frac{\left(1 - \frac{\phi}{2}\right)}{\sqrt{\phi + 3}} \cdot \frac{\sqrt{5} - 1}{4}$$

Equation 2.11

It is likely that Equation 2.11 is capable of further simplification.

The Major Pentagon Side, AB

By the Cosine Rule, the Pentagon Side AB may be defined as:-

$$AB = \sqrt{AO^2 + AO^2 - 2.AO.AO.\cos\left(\frac{2}{5}\pi\right)}$$

Equation 2.12

from which it follows that:-

$$AB = \sqrt{2.\left(\frac{1}{2}\right)^2 . \left(1 - \frac{\phi}{2}\right)} = \sqrt{\frac{1}{2}.\left(1 - \frac{\phi}{2}\right)}$$

Equation 2.13

The Circle Chord BT

Reference to Figure 2.1 permits a short chord to be imagined between Points B and T. By opposite angles, you can see that the angle BOT is π/10 radians.
Because:-

$$OB = OT = \frac{1}{2}$$

Equation 2.14

the Cosine Rule permits us to write:-

$$BT = \sqrt{OB^2 + OT^2 - 2.OB.OT.\cos\left(\frac{\pi}{10}\right)}$$

$$= \sqrt{\left(\frac{1}{2}\right)^2 + \left(\frac{1}{2}\right)^2 - 2.\frac{1}{2}.\frac{1}{2}.\frac{\sqrt{10 + 2\sqrt{5}}}{4}}$$

$$= \sqrt{\frac{1}{2} - \frac{\sqrt{\phi + 3}}{4}}$$

$$BT = \frac{\Psi}{2}\sqrt{1 - \frac{\sqrt{\phi + 3}}{2}}$$

Equation 2.15

Minor Pentagon Side FG

$$FG = 2.KG = 2.AG.\frac{\sqrt{5} - 1}{4}$$

Equation 2.16

Therefore:-

$$FG = 2.\frac{\left(1 - \frac{\phi}{2}\right)}{\sqrt{\phi + 3}}.\frac{\sqrt{5} - 1}{4}$$

Equation 2.17

giving:-

$$FG = \frac{1}{2}.\left(1 - \frac{\phi}{2}\right).\frac{(\phi^3 + 1)}{\sqrt{\phi + 3}}$$

Equation 2.18

Minor Pentagon Radius, Area and Minor Pentagon Chord

We have seen that the Minor Pentagon Radius, r, is the line segment KO whose value is $\phi/4$ (Equation 2.4).

Thusly, the Minor Pentagon Chord, KL, is yielded via the Cosine Rule as:-

$$KL = \sqrt{KO^2 + KO^2 - 2.KO.KO.\cos\left(\frac{2}{5}\pi\right)}$$

Equation 2.19

from which it follows that:-

$$KL = \sqrt{2 \cdot \left(\frac{\phi}{4}\right)^2 \cdot \left(1 - \frac{\phi}{2}\right)}$$

Equation 2.20

or alternatively, given that r = KO:-

$$KL = t = r\sqrt{\frac{5 - \sqrt{5}}{2}}$$

Equation 2.21

Areas

Area of the Minor Pentagon, a

By the Sine Rule, the Area of the Minor Pentagon FGHIJ, a, is given by:-

$$a = 5\left[\frac{1}{2} \cdot r \cdot r \cdot \sin\left(\frac{2}{5}\pi\right)\right]$$

$$= \frac{5r^2}{4}\sqrt{\frac{5 + \sqrt{5}}{2}}$$

$$= \frac{5.KO^2}{4}\sqrt{\frac{5 + (2\phi + 1)}{2}}$$

$$a = \left(\frac{5}{64}\phi^2\right)\sqrt{\phi + 3}$$

Equation 2.22

Area of the Major Pentagon, A

Since the Major Pentagon Vertex Radius is R = AO = ½ we may re-employ the Included Sine Formula for the Area of Pentagram Sector to compute Major Pentagon Area, A, as:-

$$A = 5\left[\frac{1}{2} \cdot R \cdot R \cdot \sin\left(\frac{2}{5}\pi\right)\right]$$

$$= 5\left[\frac{1}{2} \cdot \frac{1}{2} \cdot \frac{1}{2} \cdot \sin\left(\frac{2}{5}\pi\right)\right]$$

$$A = \frac{5}{8}\left[\sin\left(\frac{2}{5}\pi\right)\right]$$
Equation 2.23

and accordingly:-

$$A = \frac{5}{16}\sqrt{\phi + 3}$$
Equation 2.24

Given that R = AO = ½ the numerical value of A is 0.594410322684471

The Area A by Summation of Sections

By summing the areas of component shapes of a pentagonal sector we may compose a "non-Occam" expression of area which may nevertheless throw useful light upon several recondita of the mathematical anatomy.

The Major Pentagon Sector AOB may be resolved into four simple sections:-

(a)   The Minor Pentagon figure OKGL
(b)   The Flank Right Triangle AKG
(c)   The Flank Right Triangle GLB
(d)   The Outer Triangle AGB

Triangles AKG and GLB are congruent, and accordingly shall be computed together. Therefore:-

$$Area\ AOB = \frac{A}{5} = OKGL + 2.AKG + AGB$$
Equation 2.25

where all summands are areas.
The Area of Kite OKGL is:-

$$OKGL = \frac{KL \times OG}{2}$$
Equation 2.26

The Area of Flank Right Triangle AKG is given by numerous known formulae. We may use the formula that employs two sides and an included angle:-

$$Area = \frac{1}{2}ab\sin(\alpha)$$
Equation 2.27

But we note that the flanking triangles AKG and GLB are congruent and therefore the area of these two triangles condenses to:-

$$2AKG = \frac{AK \times KG}{2}$$
Equation 2.28

By actual reference to Equation 2.27 we may declare The Area of Outer Triangle AGB to be:-

$$AGB = \frac{1}{2} . AG . AG . \sin\left(\frac{3}{5}\pi\right)$$
Equation 2.29

Inter Alia we note that:-

$$\sin\left(\frac{3}{5}\pi\right) = \sqrt{\frac{5}{8} + \frac{\sqrt{5}}{8}} = \sqrt{\frac{1}{8}} . \sqrt{5 + \sqrt{5}}$$
Equation 2.30

Accordingly, it is possible to specify the Major Pentagon Sector AOB2 piecemeal using:-

$$AOB2 = OKGL + 2 . AKG + AGB$$
Equation 2.31

so that the Major Pentagon Area becomes:-

$$A = 5 . AOB2$$
Equation 2.32

To assist management we can move further forward to define AOB3 in these terms:-

$$AOB3 = \frac{KL \cdot OG}{2} + AK \cdot KG + \frac{1}{2} \cdot AG^2 \cdot \sin\left(\frac{3}{5}\pi\right)$$

Equation 2.33

By substitution it is now possible to expand Equation 2.31 into the Large Object (LO) displayed in full as Figure 2.2

It is often problematic to resolve Large Objects using Wolfram® Alpha® and impossible using MathCad® Express®. The old MathCad Student Edition would output a symbolic rendition as a series of multivariate polynomials that was "exact" but much longer than the input, and sadly often worse than useless. But I preferred the old MathCad, albeit that it cost me fifty pounds (nearly three hours wages at the time), over the current offering which is free but brain-dead.

But an experienced mathematician can simplify the component parts to generate a manageable and hopefully informative resolution.

You can see that all three summands are functions of ϕ.

We can condense AOB3 to the two-summand expression:-

$$AOB3 = \frac{1}{2}\left[\sqrt{2\left(\frac{\phi}{4}\right)^2\left(1 - \frac{\phi}{2}\right)} \cdot \frac{\phi}{4}\sqrt{1 + \frac{(\phi - 2)^2}{(\phi + 3)}} + \frac{\left(1 - \frac{\phi}{2}\right)^2}{\sqrt{\phi + 3}} \right.$$
$$\left. \cdot \left(\frac{\sqrt{5} - 1}{4} + \frac{1}{2}\right)\right]$$

Equation 2.34

At this stage the numerical value of AOB3 is 0.118882064536894 and Equation 2.34 may manually be simplified with, in my duffer's case, the timely assistance of Alpha®. As part of the process allow that:-

$$\sqrt{1 - \frac{\phi}{2}} = \frac{\sqrt{2 - \phi}}{\sqrt{2}}$$

Equation 2.35a

$$\phi^2 \sqrt{1 + \frac{(\phi - 2)^2}{(\phi + 3)}} = \phi^2 \sqrt{\frac{\phi^2 - 3\phi + 7}{\phi + 3}}$$

Equation 2.35b

$$\left(1 - \frac{\phi}{2}\right)^{\frac{3}{2}} \cdot \frac{(\phi + 1)}{\sqrt{\phi + 3}} = \frac{(2 - \phi)^{\frac{3}{2}} \cdot (\phi + 1)}{2\sqrt{2}.\sqrt{\phi + 3}}$$

Equation 2.35c

And also that locally:-

$$a = \phi$$

Equation 2.36a

$$b = \phi^2.\sqrt{\phi^2 - 3\phi + 7}$$

Equation 2.36b

$$c = (2 - \phi)^{\frac{3}{2}}.(\phi + 1)$$

Equation 2.36c

Noting that:-

$$AOB3 = \frac{a}{32}(b + 2c)$$

Equation 2.37

Several more steps of substitution and simplification arrive us at:-

$$AOB3 = \frac{\phi}{2^4} \cdot \left[\frac{\phi^2\sqrt{\phi^2 - 3\phi + 7}}{2} + (2 - \phi)^{\frac{3}{2}}.(\phi + 1)\right]$$

Equation 2.38

From a trigonometrical point of view you may also find it convenient to have the following form to hand:-

$$\sqrt{10 + 2\sqrt{5}} = 2\phi \left[\frac{\phi^2\sqrt{\phi^2 - 3\phi + 7}}{2} + (2 - \phi)^{\frac{3}{2}}.(\phi + 1) \right]$$

Equation 2.39

from which may be derived:-

$$\sqrt{10 + 2\sqrt{5}} = \phi^3\sqrt{\phi^2 - 3\phi + 7} + 2\sqrt{2(3\phi^2 - 6\phi + 4)}$$

Equation 2.40

At this juncture you may wish to pause to review the implications for the Major Pentagon Area A.
Area is five times AOB3:-

$$A = 5.AOB3 = \frac{5}{8} \cdot \sin\left(\frac{2}{5}\pi\right) = \frac{5}{8} \cdot \frac{\sqrt{10 + 2\sqrt{5}}}{4} = \frac{5}{16} \cdot \sqrt{\phi + 3}$$

Equation 2.41

Alternatively:-

$$AOB3 = \frac{\sqrt{\phi + 3}}{2^4}$$

Equation 2.42

from which it follows that:-

$$A = \frac{5\sqrt{\phi + 3}}{2^4}$$

Equation 2.43

The Major Pentagon Side, AB

From the Cosine Rule:-

$$AB = \sqrt{AO^2 + AO^2 - 2.AO.AO.\cos\left(\frac{2}{5}\pi\right)} = \sqrt{\frac{1}{2}\left(1 - \frac{\phi}{2}\right)}$$

Equation 2.44

$$AOB3 = \frac{\sqrt{2 \cdot \left(\frac{\phi}{4}\right)^2 \cdot \left(1 - \frac{\phi}{2}\right) \cdot \frac{\phi}{4} \cdot \sqrt{1 + \frac{(\phi - 2)^2}{(\phi + 3)}}}}{2} + \frac{1}{2} \cdot \left(1 - \frac{\phi}{2}\right) \cdot \frac{\left(\frac{1 - \phi}{2}\right)}{\sqrt{\phi + 3}} \cdot \frac{\sqrt{5} - 1}{4} + \frac{1}{2} \cdot \left(\frac{\left(\frac{1 - \phi}{2}\right)}{\sqrt{\phi + 3}}\right)^2 \cdot \sqrt{\frac{1}{8}} \cdot \sqrt{5 + \sqrt{5}} = 0.118882064536894$$

Figure 2.2
The Reductive Expression of
Major Pentagon Sector Area AOB
as a Large Algebraic Object in ϕ

Area of Triangle AFG

If $\triangle AFG$ is read as "the area of triangle AFG", etcetera, then by congruent triangles:-

$$\triangle AFG = 2.\triangle AKG$$
$$= \triangle AKG + \triangle BLG$$

Equation 2.45

The Area $\triangle AFG$, which for brevity we shall revert to AFG, is given by:-

$$AFG = 2AKG = 2 \times \frac{AK \times KG}{2}$$

Equation 2.28a

which is:-

$$AFG = 2AKG = 2 \times \frac{\frac{1}{2}\left(1 - \frac{\phi}{2}\right) \times \frac{\left(1 - \frac{\phi}{2}\right)}{\sqrt{\phi + 3}} \cdot \frac{\sqrt{5} - 1}{4}}{2}$$

Equation 2.46

Noting that:-

$$\sqrt{5} = 2\phi + 1$$

Equation 2.47

which, together with appropriate cancellations, gives:-

$$AFG = \frac{\left(1 - \frac{\phi}{2}\right)^2}{\sqrt{\phi + 3}} \cdot \frac{\phi}{4}$$

Equation 2.48

Ratios

The BT Arc/Chord Ratio

By opposite angles Angle BOT ≡ Angle KBO ≡ π/10 radians.

The Circumference of the Circle, C, is given by:-

$$C = 2R\pi = 2\frac{1}{2}\pi = \pi$$

Equation 2.49

Therefore:-

$$ArcBT = \frac{\pi}{20}$$

Equation 2.50

whilst the Chord BT is given by:-

$$ChordBT \equiv BT = \frac{\Psi}{2}\sqrt{1 - \frac{\sqrt{\phi + 3}}{2}}$$

Equation 2.15a

Accordingly the ratio ArcBT/ChordBT is:-

$$\frac{ArcBT}{ChordBT} = \frac{\pi\Psi}{20\sqrt{1 - \frac{\sqrt{\phi + 3}}{2}}}$$

Equation 2.51

The numerical value of Equation 2.51 is 1.00412420395399 and π, Ψ and φ are associated.

The Infinitesimal Circle BT Arc/Chord Ratio

Cautiously to simulate infinitesimal conditions using MathCad® Express® we may define the Infinitesimal Angle α as:-

$$\alpha = 10^{-j}.\pi$$

Equation 2.52

where j is some exponent as large as practicable without exceeding the capacity of the computing system by underflow or other technical vagaries (all of which bear highly misleading witness).

In practice, j = 6, unless you are using advanced equipment.

Now from our definitions:-

$$OB = OT = \frac{1}{2}$$

Equation 2.53

So the Infinitesimal Arc of the Major Circle, ArcBT1, is given by:-

$$ArcBT1 = \alpha R = \frac{\alpha}{2}$$

Equation 2.54

and alternatively, the identical arc is:-

$$ArcBT2 = \frac{\Psi}{2}\sqrt{\frac{\alpha^2}{2}}$$

Equation 2.55

Meanwhile, by the Cosine Rule, The Infinitesimal Chord, ChordBT1, is:-

$$ChordBT1 = \sqrt{\left(\frac{1}{2}\right)^2 + \left(\frac{1}{2}\right)^2 - 2\left(\frac{1}{2}\right)\left(\frac{1}{2}\right).\cos(\alpha)}$$

Equation 2.56

which simplifies to:-

$$ChordBT1 = \frac{\Psi}{2}\sqrt{1 - \cos(\alpha)} = \frac{\Psi}{2}\sqrt{1 - \cos(10^{-j}.\pi)}$$
$$= \frac{\Psi}{2}\sqrt{1 - \left(1 - \frac{\alpha^2}{2}\right)}$$

Equation 2.57

Therefore:-

$$\frac{ArcBT1}{ChordBT1} = \frac{ArcBT2}{ChordBT1} = \frac{\frac{\alpha}{2}}{\frac{\Psi}{2}\sqrt{1 - \left(1 - \frac{\alpha^2}{2}\right)}} = \frac{\frac{\Psi}{2}\sqrt{\frac{\alpha^2}{2}}}{\frac{\Psi}{2}\sqrt{1 - \left(1 - \frac{\alpha^2}{2}\right)}}$$
$$\approx 1$$

Equation 2.58

The optimal numerical value of this ratio computed by my 64-bit Dell® and Express® running on Windows®, and given that j = 6, is 0.999997150697196

If we wish to employ Equation 2.58 as a vehicle of association for Ψ and π we may substitute and simplify to give the following tautology:-

$$RatioBT = \frac{10^{-j}.\pi}{\Psi.\sqrt{\frac{(10^{-j}.\pi)^2}{2}}} = 1$$

Equation 2.59

which may further be simplified.

Pentagon Ratios:- Chords

Allow that the Major Pentagon Chord is T = AB whilst the Minor Pentagon Chord is t = KL.
Then:-

$$\frac{T}{t} = \frac{AB}{KL} = \Phi^3 - 1 = 3.23606797749979$$

Equation 2.60

where Φ is the Major Ratio of Phidias, which numerically is about 1.61803398874989

Pentagon Ratios:- Areas

The Minor Pentagon Area, a, is given by:-

$$a = \left(\frac{5}{64}\phi^2\right)\sqrt{\phi + 3}$$

Equation 2.22

and the Major Pentagon Area, A, by:-

$$A = \frac{5}{16}\sqrt{\phi + 3}$$

Equation 2.24

Therefore, the Pentagons' Areal Ratio is yielded by:-

$$\frac{A}{a} = (\phi^3 - 1)^2 = \phi^6 - 2\phi^3 + 1 = \frac{4}{\phi^2}$$

Equation 2.61

Circle/Major Pentagon Area Ratio

This Ratio is given by:-

$$\frac{\pi R^2}{A} = \frac{\pi}{A} \cdot \left(\frac{1}{2}\right)^2 = \frac{8\pi}{5\sqrt{10 + 2\sqrt{5}}}$$

Equation 2.62

As a function of π Equation 2.62 is transcendental.
For formality's sake name this Ratio ρ_{CMP}.
Then:-

$$\rho_{CMP} = \frac{\pi R^2}{A} = \frac{4}{5^{\frac{3}{2}}} \pi \sqrt{\Psi^2 - \phi}$$

Equation 2.63

And once more π, ϕ and Ψ stand associated and dimensionless as, perforce, all such associations must be.

The Square/Major Pentagon Area Ratio

As a convenience we may call this Ratio ρ_{SMP}. Since the width of our square PSQR is unity it must be the case that its Area is unity.
Therefore:-

$$\rho_{SMP} = \frac{1}{A} = \frac{4R^2}{A} = \frac{\Psi^8}{5\sqrt{\phi + 3}}$$

Equation 2.64

And Ψ and ϕ stand associated.

The Pentagons Chord Ratio

The Pentagons' Chord Ratio, ρ_{pent}, is:-

$$\rho_{pent} = \frac{AB}{KL} = \frac{\sqrt{\frac{1}{2} \cdot \left(1 - \frac{\phi}{2}\right)}}{\sqrt{2 \cdot \left(\frac{\phi}{4}\right)^2 \cdot \left(1 - \frac{\phi}{2}\right)}}$$

Equation 2.65

Simplifications reduce this expression to:-

$$\rho_{pent} = \sqrt{\frac{1}{4\left(\frac{\phi}{4}\right)^2}} = \Phi^3 - 1 = \phi^3 + 3\phi^2 + 3\phi$$

Equation 2.66

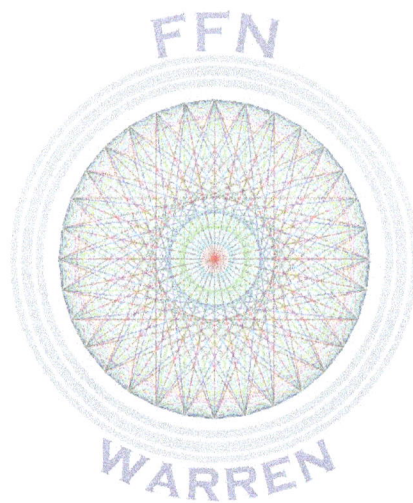

CHAPTER THREE
THE TREATMENT OF
CONVERGENT SOLUTION SERIES OF FORM
exp(γ+βk)

We shall see that when we analyse convergent solution series, especially summative series, for each of our four famous numbers we usually detect the relation:-

$$\ln(|f'|) \approx \gamma + \beta k$$
Equation 3.1

where f' (or dy/dx) is the First Differential of the (Convergent) Series of Point Solutions f(x); γ is the Linear Intercept (c_0) and β is the Linear Gradient (c_1) of the fitted curve of approximation.

Equivalently, and as we discussed in Chapter One our conjecture is that:-

$$\ln(|f'(x)|) \approx \pm U - Vx$$
Equation 1.16

where ln() is the Natural (Napierian) Logarithm function; |....| is the Absolute Value Operator; U is the Convergent Term Complexity; and V is the Logarithmatised Rate of Convergence. If the RHS negative sign is replaced by a plus in the context of any solution series then the series diverges and is useless for determining the value of a number.

The Absolute Value operation is needed because an f' value often proves negative, especially for osscilative series, and there is no logarithm of a negative number. Rendering all f' as positive does not vitiate the integrity of approximation of Equation 1.16

So our program of operations may be systematised as:-

(a) Compute the Series of First Differentials
(b) Render the First Differentials
     as Absolute Values
(c) Take the Series of Natural Logarithms of
     First Differentials
(d) Fit a Straight-line through the
     Series of Natural Logarithms
(e) Note the Intercept, Gradient and, if required, the

Correlation Coefficient of the Straight-line

Integrations of ln(|f'|)

From the above we can identify γ as U the Convergent Term Complexity; and β as V the Logarithmatised Rate of Convergence.

Employing once more our Test Polynomial with $\alpha = 14$, $\beta = 1/5$ and $\gamma = 2.333$, though α is redundant, we may specify:-

$$\ln(|f'|) = \ln\left(\left|\frac{dy}{dx}\right|\right) = \gamma + \beta k = 2.733$$

Equation 3.2

and accordingly the integral is:-

$$\int \ln(|f'|) = f' = e^{\gamma+\beta k} = 15.3789547480083$$

Equation 3.3

k is an arbitrary counter, but for our restricted needs it will not exceed two.

It follows that the Analytic Integral is:-

$$I_{ana} = \frac{e^{\gamma+\beta k}}{\beta} = 76.8947737400414$$

Equation 3.4

Somewhat counterintuitively it is more accurate to fit a quadratic to integrate Equation 3.4 than it is to fit a linear polynomial.

Therefore, for k = 0, 1, 2 we may write the Analytic Definite Integral between Lower Bound 0 and Upper Bound 2 as:-

$$\int_0^2 \frac{e^{\gamma+\beta k}}{\beta} \cdot dk = \frac{e^{\gamma+2\beta}}{\beta} - \frac{e^{\gamma+0.\beta}}{\beta} = \frac{e^{\gamma+2\beta}}{\beta} - \frac{e^{\gamma}}{\beta} = \frac{1}{\beta}\left(e^{\gamma+2\beta} - e^{\gamma}\right)$$
$$= \frac{e^{\gamma}}{\beta}\left(e^{2\beta} - 1\right)$$

Equation 3.5

or more concisely:-

$$\int_0^2 \frac{e^{\gamma+\beta k}}{\beta} \cdot dk = \frac{e^{\gamma}}{\beta}\left(e^{2\beta} - 1\right) = 25.3506654667168$$

Equation 3.6

The Constant of Integration, C, (if required) is given by:-

$$C = I_{ana} - f'$$

Equation 3.7

or:-

$$C = \frac{e^{\gamma+\beta k}}{\beta} - e^{\gamma+\beta k} = e^{\gamma+\beta k}\left(\frac{1}{\beta} - 1\right) = 61.5158189920331$$

Equation 3.8

for k = 2.

A Comparison of Two-Interval Newton-Cotes Integrations

We may reasonably think that a Trapezoidal NC numerical integral is necessary and sufficient to our straight-line model, fitting as it does straight-line segments to the series {x,y} points.

So let us define the Function of x at the three points x = 0, 1, 2 in these terms:-

$$f(0) = e^{0.\beta+\gamma}$$

Equation 3.9a

$$f(1) = e^{1.\beta+\gamma}$$

Equation 3.9b

$$f(2) = e^{2.\beta+\gamma}$$

Equation 3.9c

or generally:-

$$f(k) = e^{k.\beta+\gamma}$$

Equation 3.10

Then our First Approximation of the Trapezoidal Integral is:-

$$Trap1 = (ub - lb).\frac{1}{2}.[f(ub) + f(lb)] = (2 - 0).\frac{1}{2}[e^{2\beta+\gamma} + e^{\gamma}]$$
$$= e^{2\beta+\gamma} + e^{\gamma}$$

Equation 3.11

so that Trap1 is 25.6877764026732 and the corresponding Percentage Specific Defect is:-

$$PSD\left(\int_0^2 \frac{e^{\gamma+\beta k}}{\beta}.dk, Trap1\right) = -1.32979126878946$$

Equation 3.12

Notwithstanding that, judicious re-statement of the Trapezoidal Integral approximation to include the central data point f(1) as:-

$$Trap2 = \frac{1}{2}[f(0) + 2f(1) + f(2)]$$

$$= \frac{1}{2}[e^{\gamma} + 2e^{\beta+\gamma} + e^{2\beta+\gamma}]$$

$$= \frac{e^{\gamma}}{2} + e^{\beta+\gamma} + \frac{e^{2\beta+\gamma}}{2}$$

Equation 3.13

leads to:-

$$Trap2 = \frac{e^{\gamma}}{2}(e^{\beta} + 1)^2$$

Equation 3.14

From which:-

$$Trap2 = 25.4351114037256$$

Equation 3.15

and the Percentage Specific Defect is:-

$$PSD \left(\int_0^2 \frac{e^{\gamma+\beta k}}{\beta} . dk, Trap2 \right) = -0.33311132253988$$
Equation 3.16

Integration with Simpson's Rule

A three-point primitive Simpson's Rule integration is:-

$$Simp = \frac{1}{3}[f(0) + 4f(1) + f(2)]$$
Equation 3.17

given that our x-Interval, h, is unity.
Accordingly, elaboration of Equation 3.17 gives:-

$$Simp = \frac{1}{3}\left[e^{0.\beta+\gamma} + 4e^{1.\beta+\gamma} + e^{2.\beta+\gamma}\right]$$
$$= \frac{1}{3}\left[e^{\gamma} + 4e^{\beta+\gamma} + e^{2.\beta+\gamma}\right]$$
$$= \frac{1}{3}\left[e^{\gamma} + 4e^{1.\beta}.e^{\gamma} + e^{2.\beta}.e^{\gamma}\right]$$
Equation 3.18

or concisely:-

$$Simp = \frac{e^{\gamma}}{3}\left[1 + 4e^{\beta} + e^{2.\beta}\right]$$

Equation 3.19

Equation 3.19 may conveniently be re-expressed using the hyperbolic cosine cosh(x):-

$$Simp = \frac{2}{3}e^{\beta+\gamma}[2 + \cosh(\beta)]$$
Equation 3.20

The value of the Simpson's Rule Integration with the given data is 25.3508897374098 and:-

$$PSD\left(\int_0^2 \frac{e^{\gamma+\beta k}}{\beta} \cdot dk, Simp\right) = -0.000884673790057$$

Equation 3.21

This is strikingly nice and I dare say that a four-interval Simpson would be even closer.

CHAPTER FOUR
THE CONVERGENCE OF ESTIMATION FORMULAE:
THE LUDOLPHINE CONSTANT π

The Ludolphine Constant $\pi \approx 3.14159265358979$ has a very ancient history stretching back into the Neolithic. Attempts to award it a numerical value are at least as old, and the Ancients favored rational approximations in terms of fractions such as 22/7.

Accuracy was erratic until the Greeks recorded the theoretical underpinnings of their solution strategies.[4.1]

For example, the fraction 22/7 is said to have been offered by the mathematicians, very probably surveyors, of the Egyptian Old Kingdom which flourished circa 2400BC. The PSD of this estimate is -0.04024994347707

Ancient Babylon (c1800BC) offered 25/8, a worse figure of PSD = 0.528160567565411 whilst back in Egypt the later Rhind Papyrus (c1600BC) specified the even less accurate -0.601643040802976

More than a thousand years later the Greek genius Archimedes of Syracuse used a ninety-six-sided regular polygon to approximate a circle and, in modern terms, established:-

$$\pi_{Arch} = \frac{1}{2}\left(\frac{223}{71} + \frac{22}{7}\right) \approx 3.14185110663984$$
Equation 4.1

The PSD of this figure is -0.0082268160944

Nearly two thousand years later, around 1300AD, the great Florentine poet Dante Alighieri proposed:-

$$\pi_{Dante} = 3 + \frac{\sqrt{2}}{10} = 3.14142135623731$$
Equation 4.2

a slightly superior estimate with a PSD of 0.005452564077259

Also around the 14th Century AD, the Indian mathematician Madhava made real progress in the estimation of π by deriving the approximation Pi8 as:-

$$Pi8 = \sqrt{12} \sum_{k=0}^{n} \frac{(-3)^{-k}}{2k+1}$$

Equation 4.3

In this formula, MathCad® Express® will not cope with n>29, but using that value we achieve 3.14159265358979

The resulting Madhava PSD is -0.000000000000014, and we have to hand a π value sufficient to all but the most exacting applications.

Modern Developments

The Euler Series (18th Century)

The Euler Series for Pi is due to the Russian-based Swiss mathematician Leonhard Euler and is:-

$$\pi_{Euler} = \sqrt{6} \sum_{k=1}^{n} \frac{1}{k^2}$$

Equation 4.4

where n = 48, the practical limit for 64-bit MathCad. PSD(π_{fido},π_{Euler}) is 0.628682982575791

The Ramanujan-Sato Series (20th Century)

The Ramanujan-Sato Series generates an estimate of the *reciprocal of* π that we shall call RePi14.

At this juncture, it is convenient to define:-

A = 26390
B = 1103
C = 396

and:-

$$s(k) = \frac{(4k)!}{(k!)^4}$$

Equation 4.5

For a mere m = 2 we get the following result:-

$$\frac{1}{\pi} = RePi4 \approx \frac{2\sqrt{2}}{99^2}\left[\sum_{k=0}^{m} s(k) \cdot \frac{Ak + B}{C^{4k}}\right] = 0.318309886183791$$

Equation 4.6

by which the Ramanujan-Sato π Estimate, π_{RS}, is 3.14159265358979

The m = 2 Ramanujan-Sato PSD($1/\pi_{fido}$,RePi4) is zero.

The formula seems relatively involved, but the result speaks for itself.

The Chudrovsky Formula (1987)

The Chudrovsky Formula also defines π in terms of its reciprocal.

Allow that RePi15 is the Reciprocal of π as estimated by the Chudrovsky Formula.

Then:-

$$RePi15 = \frac{\sqrt{10005}}{4270934400}\sum_{k=0}^{m}(6k)! \cdot \left[\frac{13591409 + 545140134k}{(3k)!\,(k!)^3(-640320)^{3k}}\right]$$
$$= 0.318309886183791$$

Equation 4.7

The Chudrovsky PSD($1/\pi_{fido}$,RePi15) is 0.000000000000017

The BBP Digit-Extraction Spigot Algorithm (Plouffe 1995)

The BBP Digit-Extraction Spigot Algorithm is one of a family of formulae for the estimation of irrational number values which may be generalised as:-

$$Est = \sum_{k=0}^{m} \frac{i}{j} \sum_{l=1}^{n} \left[\pm \frac{u_l}{w.k + v_l} \right]$$

Equation 4.8

where Est is an Estimated Value of a Specific Number; m is the number of Serial Iterations (researcher-determined); n is the Number of Inner Cycles (four is typical, but six and other low counts are feasible); w is typically 4, 6 or 8; and u_l and v_l are Coefficients.

i, j, k, l, m, n, u_l, v_l and w are all integers.

BBP-type spigot algorithms are as capable of computing whole estimates as they are at defining intra-numeric digits.

The original BBP Digit-Extraction Spigot Algorithm for π is:-

$$Pi16 = \sum_{k=0}^{m} \frac{1}{16^k} \left(\frac{4}{8k + 1} - \frac{2}{8k + 4} - \frac{1}{8k + 5} - \frac{1}{8k + 6} \right)$$

$$= 3.14159265358979$$

Equation 4.9

As you can see this renders whole-pi in denary.

PSD(π_{fido},Pi16) is zero to MathCad® fifteen-figures of display.

A Comparison of Convergent Series for Pi

For five of the more effective and extensible methods discussed above I computed solution series according to the equation:-

$$\ln(|f'|) \approx \gamma + \beta.k$$

Equation 4.10

The Euler Method is shown in Figure 4.1 by the blue dashed line which is closely shadowed by its linear regression trendline as sensibly coincident gray dots.

In agreement with its poor PSD result, the Euler Method is markedly superior in terms of Complexity, $\gamma = -4.16620018$, but very inferior in terms of Convergence Rate, $\beta = -0.16661613$. $R^2 = 0.85845297$ which betrays the poor fit of a straight line to the trace of Euler convergence, *but a higher-degree regression is even less meaningful*.

At the other end of efficacy, the Chudrovski Formula is of high Complexity, $\gamma = +80.04280913$, but of very precipitate Convergence, $\beta = -32.70894077$

In broad terms, the lower the absolute γ the better; the more negative the β the better. For convergence β must always be negative. $R^2 = 0.99999700$ which shows the excellent fit of a straight line to the Chudrovski trace: Perfection is represented by $R^2 = 1$.

On Figure 4.1 the Chudrovski line is in solid red.

The Ramanujan-Sato method plotted as the light orange long dashes with the regression line as dots is also good with Complexity γ half that of Chudrovski and Convergence Rate β about two-thirds that of Chudrovsky.

Madrava (green) and BBP (yellow) methods are of relatively low efficacy in intermediate positions.

The correlation of PSD and Equation 4.9 parameters remains to be explored.

As in life, you get what you pay for and better performance is purchased with greater complexity, despite what theoreticians may tell you.

Summaries of Results

Table 4.1 summarises EXCEL® summative and linear regression results for five of the serial structures that exhibited reasonable rates of convergence. Table 4.1 confirms that all five methods converge to π with more or less accuracy, but unshown is the Number of Iterations it takes to get to "acceptable" π in each case (i.e. about thirteen digits).

For your convenience, but for illustrative purposes only, the requisite Iterations, k, for "acceptable" π are:-

	k
Euler	>48
Madrava	24
BBP	11
Ramanujan-Sato	3
Chudrovski	2

The table shows summary statistics for the Natural Logarithms of the Absolute First Differentials around convergent points. Importantly, the Table lists Intercept (Complexity) γ; Gradient (Rate of Convergence) β; and Coefficient of Determination, R^2, for each of the five methods as given on the plot Figure 4.1

Students should note that *any* algebraic polynomial fitment to these systems is *arbitrary* and is of descriptive importance only: So elaborate attempts statistically to fit octic equations or even quadratics is from the point of view of theory a waste of time. This is not to rule out their usefulness in the clarification of thought or the inspiration of more mathematically-sound treatments.

Not to mention industrial utility...

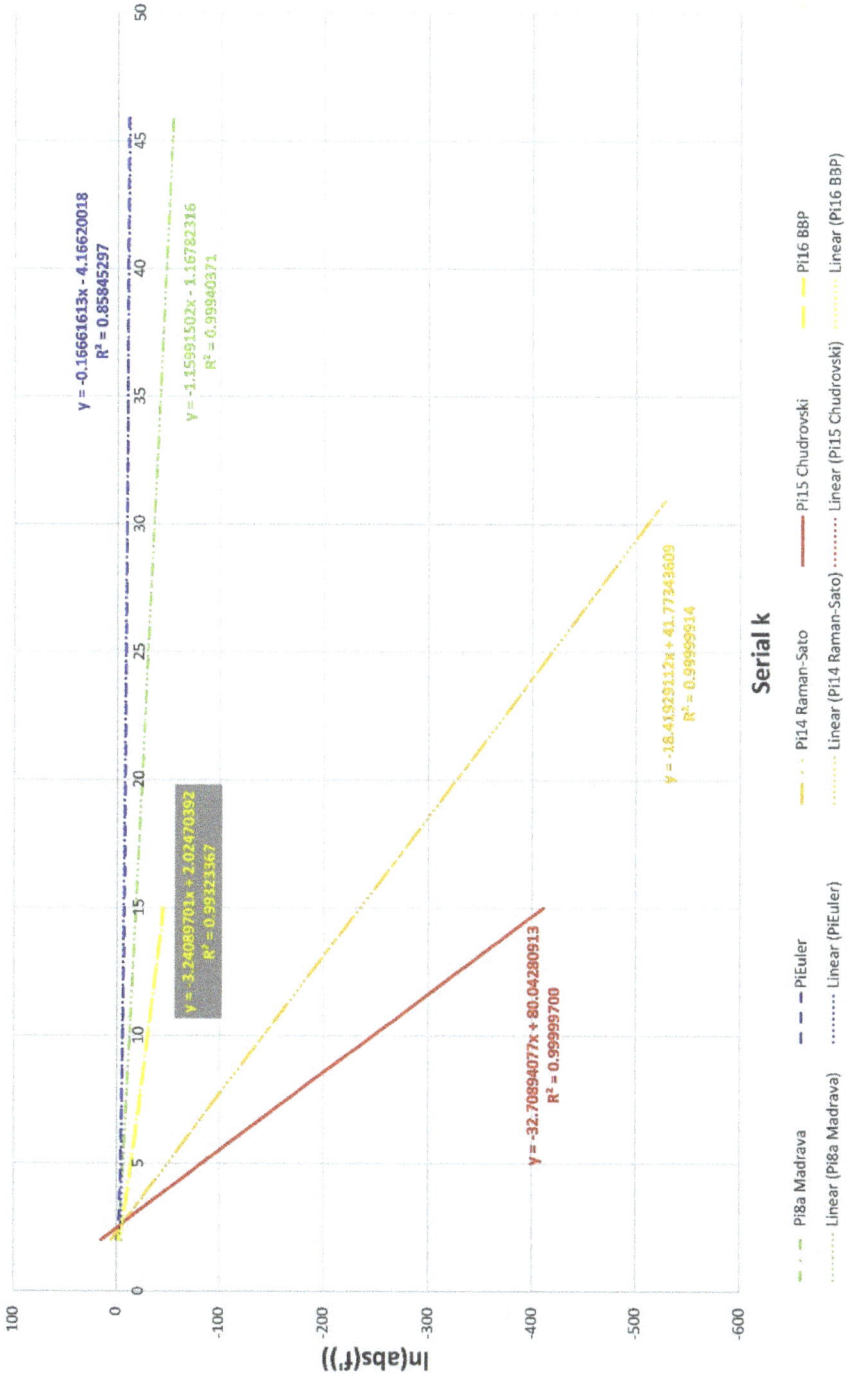

Figure 4.1
Convergent Algorithms for π
ln(|f'|) versus Number of Iterations, k

Fiducial Values

The Ludolphine Constant	π	3.14159265358979
The Pythagoras Constant	Ψ	1.41421356237310
The (Major) Ratio of Phidias	Φ	1.61803398874989
The Euler Constant	e	2.71828182845905

CONVERGENTS

	Madhava	Euler	Ramanujan-Sato	Chudrovski	BBP Spigot Algorithm
	Pi8a	PiEuler	Pi14	Pi15	Pi16
Sum or Product	0.9068996821171109	1.6243162404933670	1103.00026831970000	13591408.99999970000000	3.141592653589790
Multiplier	3.4641016151377750	2.4494897427831800	0.0002888565565223	0.000000023419933	1.000000000000000
Result	3.141592653589790	3.12184195194820	3.141592653589790	3.141592653589790	3.141592653589790
PSD(fido, result)	-0.000000000000028	0.628682982575791	0.000000000000000	-0.000000000000014	0.000000000000000
Linear Regression Parameters for the Count k (x) versus ln(|f|) (y)					
Intercept	-1.167823159636950	-4.166200184223430	41.773436088468900	80.042809131108300	2.024703918867470
Grade	-1.159915018409610	-0.166616133696459	-18.419291124860200	-32.708940769677100	-3.240897006319550
RSQ	0.999403713666952	0.858452970085900	0.999999136969575	0.999997004400978	0.993233674388125

Table 4.1
Summary Results for π Selected Approximations

CHAPTER FIVE
THE CONVERGENCE OF ESTIMATION FORMULAE:
THE PYTHAGORAS' CONSTANT Ψ

Readers will often find it convenient when referring to mathematical functions to utilise Abramowitz and Stegun[5.1] which though old is intelligible and thoroughly comprehensive, if you will forgive the pleonasm.

Also bear in mind Appendix A of this book "Some Convenient Functions".

The Square Root of Two is:-

$$\Psi = \sqrt{2} = 1.4142135623731$$
Equation 5.1

and interest in, and numerical approximations of this root, date back at least to the Iron Age in Europe.

The value of Equation 5.1 will be adopted as the fiducial approximation of Ψ, Ψ_{fido}[5.2,5.3].

Rational Approximations

The Pythagoras' Constant, Ψ, is an irrational number but rational approximations of sundry qualities are computable.

For example, allow the Quadratic Coefficients, α, β and γ as below:-

$$\alpha = c_2 = 14$$
Equation 5.2a

$$\beta = c_1 = \frac{1}{5}$$

Equation 5.2b

$$\gamma = c_0 = \frac{7}{3}$$
Equation 5.2c

Then:-

$$\Psi_{rat1} = -1 + \frac{1}{1000}\int_0^8 \alpha x^2 + \beta x + c.\,dx = 1.4144$$
Equation 5.3

The Percentage Specific Defect is:-

$$PSD\left(\Psi_{fido}, \Psi_{rat1}\right) = -0.013183131025282$$
Equation 5.4

The Babylonian Approximation[5.4]

This extensible series takes the form:-

$$\Psi0 = 1 + \frac{24}{60} + \frac{51}{60^2} + \frac{10}{60^3} = 1.41421296296296$$
Equation 5.5

The Percentage Specific Defect is:-

$$PSD\left(\Psi_{fido}, \Psi0\right) = 0.000042384696919$$
Equation 5.6

The Extended Babylonian Method[5.5,5.6]

Define the array of integer Term Coefficients β_k as:-

$\beta_0 = 1$	$\beta_1 = 24$	$\beta_2 = 51$	$\beta_3 = 10$
$\beta_4 = 7$	$\beta_5 = 46$	$\beta_6 = 6$	$\beta_7 = 4$
$\beta_8 = 44$	$\beta_9 = 50$		

Then:-

$$\Psi0a = \sum_{k=0}^{m3} \frac{\beta_k}{60^k} = 1.4142135623731$$
Equation 5.7

and:-

$$PSD\left(\Psi_{fido,}\Psi0a\right) = 0$$
Equation 5.8

The necessary and sufficient m3 is nine.

<u>Egyptian Fractions</u>

For Egyptian Fractions the n summation limit is 48, and further allow that:-

$$a_0 = 0$$
Equation 5.9a

$$a_1 = 6$$
Equation 5.9b

$$a_i = 34a_{i-1} - a_{i-2}$$
Equation 5.9c

Then:-

$$\Psi7 = \frac{3}{2} - \frac{1}{2} \cdot \sum_{i=1}^{n} \frac{1}{a_i}$$
Equation 5.10

and:-

$$PSD\left(\Psi_{fido,}\Psi7\right) = 0.005106335460268$$
Equation 5.11

<u>Trigonometric Products</u>

In each of these three trigonometric product series n = 48

Product #1

$$\Psi inv1 = \prod_{k=0}^{n}\left[1 - \frac{1}{(4k+2)^2}\right]$$
Equation 5.12

and thus:-

$$\Psi 1 = \frac{1}{\Psi inv1} = 1.41241092316924$$
Equation 5.13

whilst:-

$$PSD\left(\Psi_{fido,}\Psi 1\right) = 0.127465840507752$$
Equation 5.14

Product #2

$$\Psi 2 = \prod_{k=0}^{n}\frac{(4k+2)^2}{(4k+1)(4k+3)} = 1.41241092316924$$
Equation 5.15

and:-

$$PSD\left(\Psi_{fido,}\Psi 2\right) = 0.127465840507752$$
Equation 5.16

Equations 5.13 and 5.15 are identities.

Product #3

$$\Psi 3 = \prod_{k=0}^{n}\left(1 - \frac{1}{4k+1}\right)\left(1 - \frac{1}{4k+3}\right) = 1.41241092316924$$
Equation 5.17

and:-

$$PSD\left(\Psi_{fido,}\Psi 3\right) = 0.127465840507579$$
Equation 5.18

There is considerable evidence that Equations 5.13, 5.15 and 5.17 are identities despite discrepancies in the last three digits of the Product #2 and Product #3 PSDs.

Taylor Series Trigonometric Summation

$$\Psi inv4 = \sum_{k=0}^{8}\frac{(-1)^k\left(\frac{\pi}{4}\right)^{2k}}{(2k)!} = 0.707106781186547$$
Equation 5.19

Hence:-

$$\Psi 4 = \frac{1}{\Psi inv4} = 1.4142135623731$$
Equation 5.20

The PSD is given as:-

$$PSD\left(\Psi_{fido,}\Psi 4\right) = 0$$
Equation 5.21

Taylor Series of sqrt(1+x) with x = 1

$$\Psi 5 = \frac{3}{2} + \sum_{i=1}^{48}(-1)^i \cdot \frac{\prod_{j=0}^{i-1}(2j+1)}{\prod_{j=0}^{i+1}(2j)} = 1.41462158647635$$
Equation 5.22

The PSD is:-

$$PSD(\Psi_{fido,}\Psi 5) = -0.028851661030178$$
Equation 5.23

Despite its term complexity the Ψ5 estimate of Equation 5.22 is markedly inferior to the Ψ4 estimate of Equations 5.19 and 5.20

Taylor Series of sqrt(1+x) with x = 1
Accelerated by the Euler Transform

$$\Psi 6 = \sum_{k=0}^{48} \frac{(2k+1)!}{2^{3k+1}.(k!)^2} = 1.41421356237308$$
Equation 5.24

which yields the PSD:-

$$PSD(\Psi_{fido,}\Psi 6) = 0.000000000001021$$
Equation 5.25

Of these three classical methods involving the Taylor Series the first Ψ4 involving Equations 5.19 and 5.20 is superior.

Weisstein's BBP-type Formula[5.7]

The Weisstein Formula for Ψ (Ψ8) is "exact" to fifteen-figures using MathCad® Express®. The formula is:-

$$\Psi 8 = \frac{1}{\pi}\sum_{k=0}^{15} \frac{1}{(-8)^k}\left(\frac{4}{6k+1} + \frac{1}{6k+3} + \frac{1}{6k+5}\right)$$
$$= 1.4142135623731$$
Equation 5.26

which yields the PSD:-

$$PSD(\Psi_{fido,}\Psi 8) = 0$$
Equation 5.27

A Comparison of Convergent Series for Psi

For six of the more effective and extensible methods discussed above I computed solution series according to the equation:-

$$\ln(|f'|) \approx \gamma + \beta.k$$
Equation 5.28

These six methods, depicted in the plot Figure 5.1, are:-

Ψ0a   Extended Babylonian Method
       Light Orange Dashes
Ψ7   Egyptian Fractions
       Yellow Dashes
Ψ4   TS Simple Trig Summation
       Red Continuous Line
Ψ6   Accelerated TS of Trig Summation
       Blue Short Dashes
Ψ1   Trigonometric Product #1
       Purple Continuous Line
Ψ8   Weisstein's BBP-type Formula
       Gray Dashes

The Trigonometric Product #1 Ψ1 is shown in Figure 5.1 by the unsteady purple dashed line which is roughly shadowed by its linear regression trendline as indistinct purple dots.

In agreement with its poor PSD result, the Trigonometric Product #1 Ψ1 is superior in terms of Complexity, γ = -7.29267972, but very inferior in terms of Convergence Rate, β = -0.15722681. The Coefficient R^2 = 0.89424739 which betrays the poor fit of a straight line to the trace of Trigonometric Product #1 Ψ1.

At the other end of efficacy, the Taylor Series Simple Trig Summation Ψ4 is of high Complexity, γ = -31.941, but of very precipitate Convergence, β = -7.7895. The linear R^2 = 0.9926, but the quadratic R^2 = 0.99955450 shows that a second-degree fitment is only marginally better, and of less certain interpretability.

In broad terms, the lower the absolute γ the better; the more negative the β the better. For convergence β must always be negative. Ideally, R^2 = 1, but an R^2 = 0.9926 is more than desirable.

On Figure 5.1 the Taylor Series Simple Trigonometric Summation Ψ4 line is in solid red.

The correlation of PSD and Equation 5.28 parameters remains to be explored.

Summaries of Results

Table 5.1 summarises EXCEL® summative and linear regression results for eight of the serial structures that exhibited reasonable rates of convergence. Table 5.1 confirms that all eight methods converge to Ψ with more or less accuracy, but unshown is the Number of Iterations it takes to get to "acceptable" Ψ in each case (i.e. about thirteen digits).

For your convenience, but for illustrative purposes only, the requisite Iterations, k, for "acceptable" Ψ in the best six of the methods are:-

		k
Taylor Series Simple Trig Summation	Ψ4	9
Extended Babylonian Method	Ψ0α	15
Egyptian Fractions (Extended)	Ψ7	15
Weisstein's BBP-type Formula	Ψ8	16
Accelerated TS of Trig Summation	Ψ6	48
Trigonometric Product #1	Ψ1	48

Table 5.1 lists Intercept (Complexity) γ; Gradient (Rate of Convergence) β; and Coefficient of Determination, R^2, for each of the eight methods as given on the plot Figure 5.1

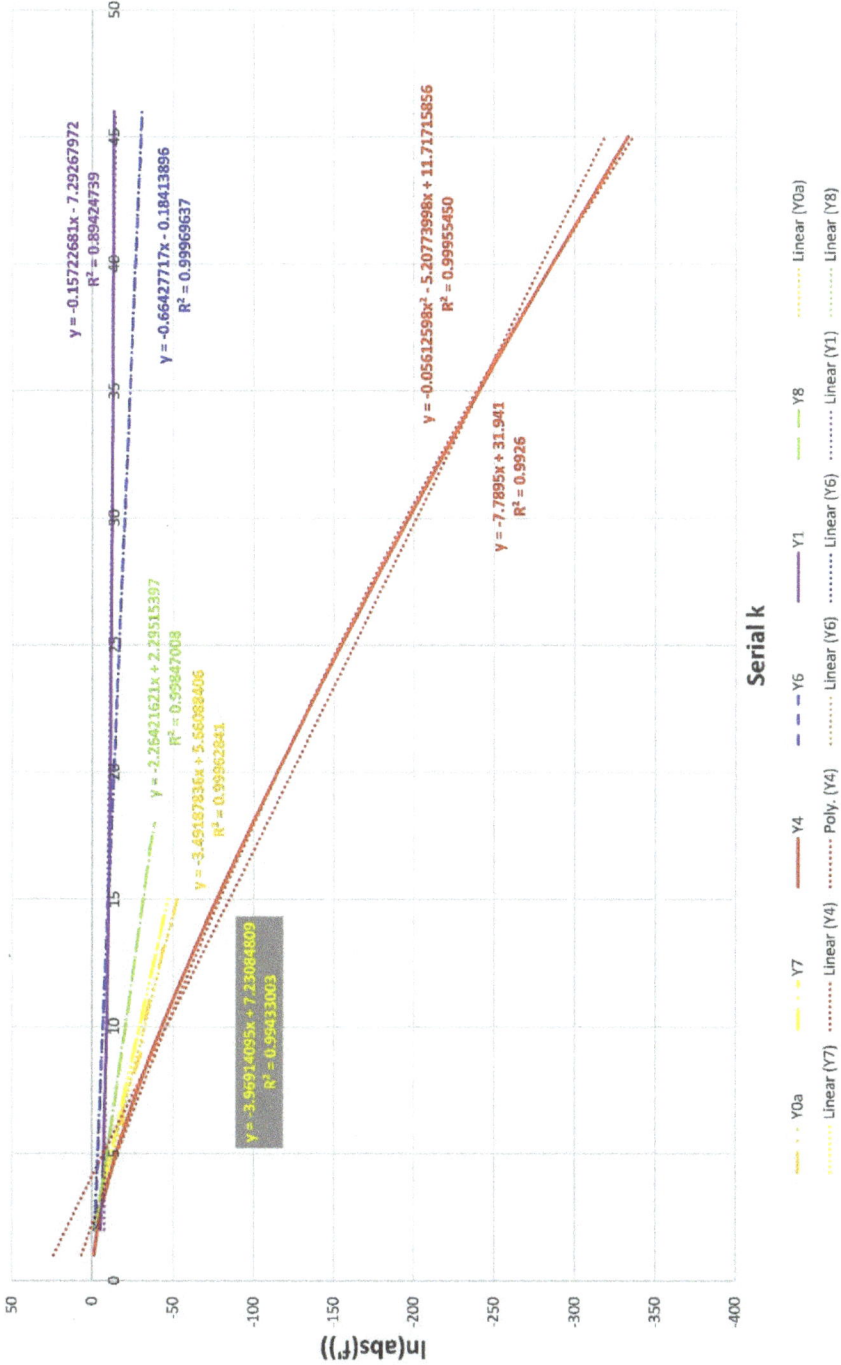

Figure 5.1
Convergent Algorithms for Ψ
ln(|f'|) versus Number of Iterations, k

Fiducial Values

The Ludolphine Constant	π	3.14159265358979
The Pythagoras Constant	Ψ	1.41421356237310
The (Major) Ratio of Phidias	Φ	1.61803398874989
The Euler Constant	e	2.71828182845905

CONVERGENTS

	Extended Babylonian Approximation	Egyptian Approximation	Taylor Series Trigonometric Summation	Taylor Series of SQRT(1+x) with x = 1 Accelerated with the Euler Transform	Trig Product #1	Trig Product #2	Trig Product #3	Weisstein's BBP-type Formula
Sum or Product	1.41421356237310	0.171717304231048	0.70706781286547	1.41421356237308	0.70809251129372	1.41241092316924	1.41241092316924	4.44288293815837
Multiplier	1.00000000000000	1.41414134778480	1.41421356237310	1.41421356237308	1.41241092316924	1.41241092316924	1.41241092316924	0.31830986183791
Result	1.41421356237310	1.41414134778480	0.00000000000000	0.00000000000001021	0.127465840507752	0.127465840507752	0.127465840507759	1.41421356237310
PSD(fido, result)	0.00000000000000	0.00651063354460268	0.00000000000000	0.00000000000000000				0.00000000000000000
	ΨQa	Ψ7	Ψ4	Ψ6	Ψ1	Ψ2	Ψ3	Ψ8

Linear Regression Parameters for the Count k (x) versus $\ln(|\Psi|)$ (y)

	ΨQa	Ψ7	Ψ4	Ψ6	Ψ1	Ψ2	Ψ3	Ψ8
Intercept	7.230848089021930	5.660884058408590	39.730754925540700	-0.184138964466360	-7.292679722506620	-7.219686371666320	-7.219686371668590	2.295153956892660
Grade	-3.969140945758940	-3.451878364557820	-7.789535040070250	-0.664277168207107	-0.157226813206622	-0.159524524050871	-0.159524524050755	-2.264216200917680
RSQ	0.994330028773808	0.999628410482425	0.992651131408094	0.999696637792928	0.894247394482847	0.894352000191120	0.879435200018530	0.998470076763613

Table 5.1
Summary Results for Ψ Selected Approximations

CHAPTER SIX
THE CONVERGENCE OF ESTIMATION FORMULAE:
EULER'S NUMBER e

Leonhard Euler started to use the letter *e* for this constant in 1727 or 1728, in an unpublished paper on explosive forces in cannons, and in a letter to Christian Goldbach on 25 November 1731.[6.1,6.2]

1727AD was the year that Issac Newton died but his methods were already well-known, and indeed the concept of the Natural Logarithm had been floating around in the intellectual aether since the work of the Scottish mathematician John Napier in the first years of the seventeenth-century.

You will find Euler's Number referred to as the Base of Natural Logarithms, Napier's Constant, or, especially in my past work, the Napierian Base. Other arbitrary names are likely to be seen, but caution is advisable because, Euler's Constant, etcetera, refers to something different.

Euler's Number is transcendental.

Euler's Number is by definition:-

$$e = \lim_{n \to \infty} \left(1 + \frac{1}{n}\right)^n \approx 2.71828182845904523536$$
Equation 6.1

It is also the case that:-

$$e = \lim_{n \to \infty} \left(\frac{(n+1)^{n+1}}{n^n} - \frac{n^n}{(n-1)^{n-1}}\right)$$
$$\approx 2.71828182845904523536$$
Equation 6.2

<u>Brothers Compressed Newton Series #2</u>[6.3]

        The Newton Series for e comes in a variety of flavors. Some are efficient: Some are less so. Harlan J Brothers reviews a number of these variations in his 2004 paper.

eB2

        For eB2 allow that m3 = 8 iterations. Then:-

$$eB2 = \sum_{k=0}^{m3} \frac{2k + 2}{(2k + 1)!} = 2.71828182845904$$

Equation 6.3

PSD(e,eB2) is 0.000000000000016

eB11

        For eB11 allow that m3 = 6 iterations. Then:-

$$eB11 = \sum_{k=0}^{m3} \frac{(3k)^2 + 1}{(3k)!} = 2.71828182845905$$

Equation 6.4

PSD(e,eB11) is -0.000000000000016

eB2eB11

        The PSDs of eB2 and eB11 balance, suggesting that the arithmetic average will more accurately represent Euler's Number. Accordingly:-

$$eB2eB11 = \frac{1}{2}\left(\sum_{k=0}^{8} \frac{2k + 2}{(2k + 1)!} + \sum_{k=0}^{6} \frac{(3k)^2 + 1}{(3k)!} \right)$$

Equation 6.5

PSD(e,eB2eB11) is zero.

Four Famous Numbers   Page 80 of 210   James R Warren

Powering

In his 2004 paper Brothers make the point that the accuracy of certain Newton Series formulations, $f(x)$, can be enhanced by "powering" such that if p is the Power of the Series Summation and x is the Reciprocal of that power then:-

$$e^x = f(x)$$

Equation 6.6

and:-

$$e^{px} = [f(x)]^p$$

Equation 6.7

For example, in "squaring" p = 2. Hence:-

$$e^2 = eB14 = \sum_{k=0}^{m3} \frac{x^{2k}(x + 2k + 1)}{(2k + 1)!}$$

Equation 6.8

This series is rapidly-convergent, and it is only necessary that m3 = 7.

It follows from Equation 6.8 that:-

$$f(x) = 1.64872127070013$$

Equation 6.9

and:-

$$e^{p.x} = e^{2.\frac{1}{2}} = \left[\sum_{k=0}^{m3} \frac{x^{2k}(x + 2k + 1)}{(2k + 1)!}\right]^p = 2.71828182845905$$

Equation 6.10

for which the PSD(e,eB142) is -0.000000000000016.

An alternative formulation of Equation 6.8 is:-

$$eB14root = \left[\sum_{k=0}^{m3} \frac{\left(\frac{1}{2}\right)^{2k}\left(2k+\frac{3}{2}\right)}{(2k+1)}\right]^2 = \left[\sum_{k=0}^{m3} \frac{(2)^{-2k}\left(2k+\frac{3}{2}\right)}{(2k+1)}\right]^2$$

Equation 6.11

but this offers no detectible advantage over Equation 6.8

Theoretically, the accuracy of estimation increases with increase in the integer p, but in practice p = 16 gives a PSD of - 0.000000000000018 and p = 64 yields a PSD value of 0.000000000000131

A Comparison of Brothers-type Convergent Series for e

For six of the Brothers methods discussed above I computed solution series according to the equation:-

$$\ln(|f'|) \approx \alpha k^2 + \beta.k + \gamma$$
Equation 6.12

For none of the Brothers methods was it realistic to retain the linear model used with the first differentials of the convergent involving π or Ψ.

A quadratic model according to Equation 6.12 gave a satisfactory fit in every case.

eB2, eB11 and their arithmetic mean eB2EB11 gave the least satisfactory convergence rates (for fifteen-digit accuracy), whist the "powered" arrangements eB14p2 (p = 2), eB14p16 (p = 16), and eB14p64 (p = 64) are increasingly convergent, and if you will forgive the phrase, increasingly linear.

These six methods are, depicted in the plot Figure 6.1 and summary statistics in Table 6.1

For your convenience, but for illustrative purposes only, the requisite Iterations, k, for "acceptable" e in the best six of the methods are:-

	k
eB14p64	4
eB14p16	5
eB14root	8

eB14p2	8
eB2eB11 (Compost 1)	9
eB11	7
eB2	9

<u>The Accelerated Wallis Product (Guillera and Sondow, 2005, Sondow 2005)[6.4]</u>

This product takes the form:-

$$WGS = \frac{\pi}{2} = e^{\sum_{n=1}^{m3} \frac{1}{2^n} \sum_{k=0}^{n} (-1)^{k+1} \binom{n}{k} . \ln(k+1)}$$

Equation 6.13

or:-

$$2 \times WGS = \pi = 2e^{\sum_{n=1}^{m3} \frac{1}{2^n} \sum_{k=0}^{n} (-1)^{k+1} \binom{n}{k} . \ln(k+1)}$$

Equation 6.14

PSD(π,2.WGS) is 0.000000000000085
m3 is adequate at 48.
In other terms we may state:-

$$s = \sum_{n=1}^{m3} \frac{1}{2^n} \sum_{k=0}^{n} (-1)^{k+1} \binom{n}{k} . \ln(k+1)$$

Equation 6.15

so that:-

$$\pi = 2e^s$$

Equation 6.16

It follows that:-

$$e_{WGS} = exp\left[\frac{1}{s} . \ln\left(\frac{\pi}{2}\right)\right]$$

Equation 6.17

Where e_{WGS} is the Wallis Series Estimate of e.

PSD(e,ewGS) is -0.000000000000212

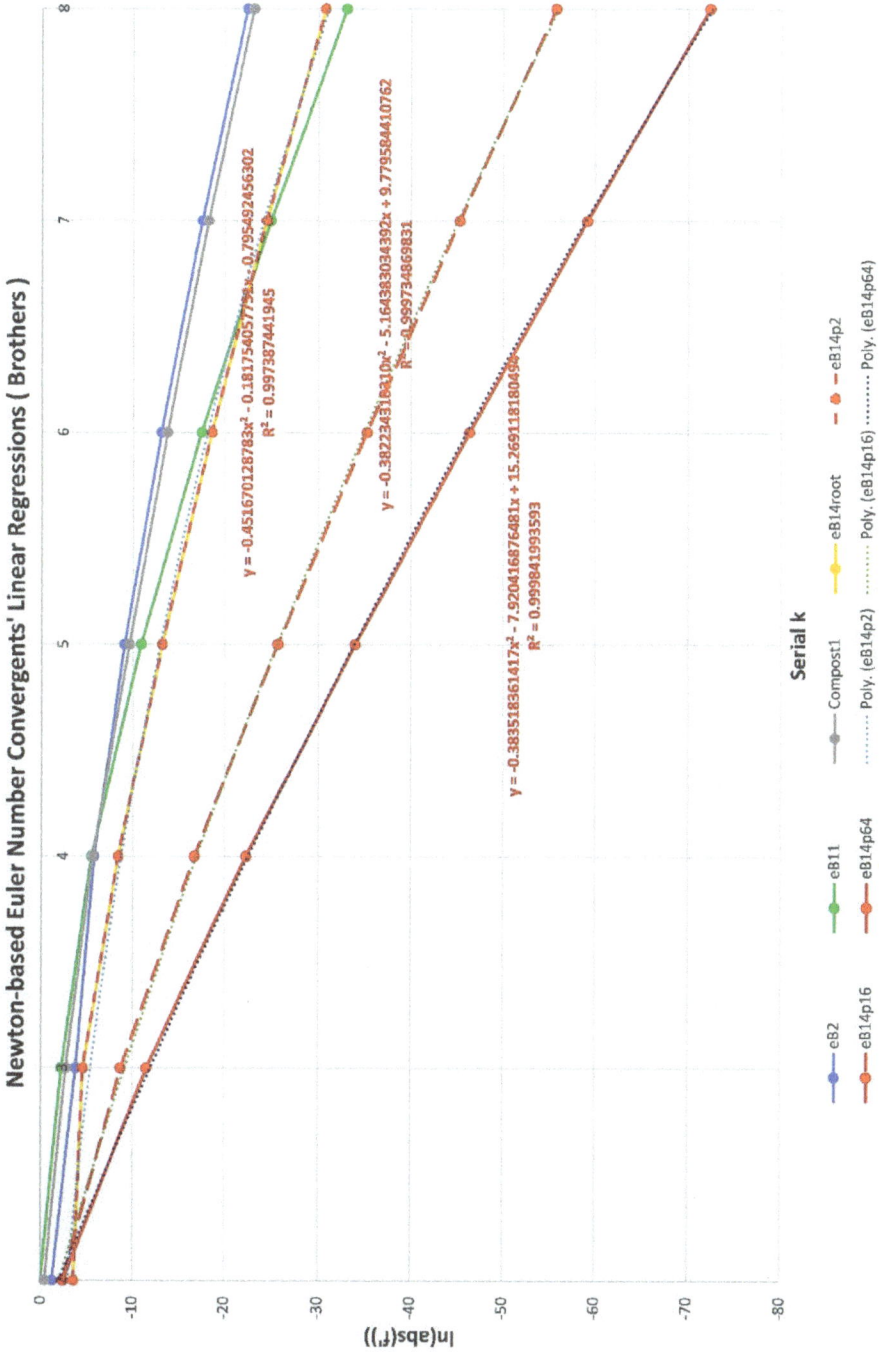

Figure 6.1
Brothers-type Convergent Algorithms for e
ln(|f'|) versus Number of Iterations, k

Fiducial Values

The Ludolphine Constant	π	3.1415926535 8979
The Pythagoras Constant	ψ	1.4142135623 7310
The (Major) Ratio of Phidias	Φ	1.6180339887 4989
The Euler Constant	e	2.7182818284 5905

CONVERGENTS

	Brother's Compressed Newton Series Number 2	Brother's Compressed Newton Series Number 11	Mean of eBrothers2 and eBrothers11	Brother's Root Algorithm	Powering p 2 x 0.5 Brother's Powering Formula Corrector Exponent 2	Powering p 16 x 0.0625 Brother's Powering Formula Corrector Exponent 16	Powering p 64 x 0.015625 Brother's Powering Formula Corrector Exponent 64
	eB2	eB11	Compost1	eB14root	eB14p2	eB14p16	eB14p64
Sum or Product	2.7182818284 59040	2.7182818284 59050	2.7182818284 59050	1.6487212707 00130	1.6487212707 00130	1.0644944589 17860	1.0157477085 86690
Corrector	1.0000000000 00000	1.0000000000 00000	1.0000000000 00000	1.0000000000 00000	0.0000000000 00000	0.0000000000 00000	0.0000000000 00000
Result	2.7182818284 59040	2.7182818284 59040	2.7182818284 59050	2.7182818284 59050	2.7182818284 59050	2.7182818284 59050	2.7182818284 59030
PSD(fido, result)	0.0000000000 000016	-0.0000000000 000016	-0.0000000000 000016	-0.0000000000 000016	-0.0000000000 000016	-0.0000000000 000180	0.0000000000 000441

Linear Regression Parameters for the Count k (x) versus $\ln(|F|)$ (y)

	eB2	eB11	Compost1	eB14root	eB14p2	eB14p16	eB14p64
Intercept	7.119742508	14.54726216	8.642145479	8.689580248	8.689580248	17.80650493	23.32300377
Grade	-3.510413523	-5.600131496	-3.830632183	-4.698455346	-4.698455346	-8.986726137	-11.75560049
RSQ	0.976872626	0.962959468	0.986703471	0.970481931	0.970481931	0.994338375	0.996659623

Quadratic Regression Parameters for the Count k (x) versus $\ln(|F|)$ (y)

	eB14p2	eB14p16	eB14p64
c_0	-0.7954492456	9.779584441	15.26911818
c_1	-0.1817540058	-5.16438303	-7.920416876
c_2	-0.451670129	-0.38223431	-0.383518361
RSQ	0.997387442	0.99973487	0.999841994

Table 6.1
Summary Results for e Selected Approximations

Other Products for e

As a general statement, product series tend to be complicated and computationally unsteady. Nevertheless, they may be useful from the points of view of insight and development and accordingly I offer a further selection.

Pippenger's Product (1980)[6.5]

Allow that Pippenger's Product approximation for e is ewiki13 given by:-

$$ewiki13 \approx 2.\left(\frac{2}{1}\right)^{\frac{1}{2}} \cdot \left(\frac{2}{3}\cdot\frac{4}{3}\right)^{\frac{1}{4}} \cdot \left(\frac{4}{5}\cdot\frac{6}{5}\cdot\frac{6}{7}\cdot\frac{8}{7}\right)^{\frac{1}{8}}$$

Equation 6.18

The PSD(e,ewiki13) is -0.259822125200446

The Exponential Function as an Infinite Product[6.6]

Using MathCad® I found that the Number of Product Terms n for "adequate" (i.e. 13-figure) solution was eighteen (n = 18). Therefore, given that:-

$$f_1 = 1$$

Equation 6.19a

$$f_{k+1} = (k + 1).(f_k + 1)$$

Equation 6.19b

then:-

$$eStack16 = \prod_{k=1}^{n} \frac{f_k + 1}{f_k} = 2.71828182845904$$

Equation 6.20

PSD(e_fido,eStack16) = 0.000000000000049

A Product involving the Möbius Function

The Möbius Function μ_k can take one of three integer values, conventionally -1, 0 or +1.

The first twenty-four values of the Möbius Function for "adequate" e are listed in Table 6.2:-

k	1	2	3	4	5	6	7	8	9	10	11	12
μ_k	1	-1	-1	0	-1	1	-1	0	0	1	-1	0

k	13	14	15	16	17	18	19	20	21	22	23	24
μ_k	-1	1	1	0	-1	0	-1	0	1	1	-1	0

Table 6.2
The First Twenty-Four Values of the
Möbius Function

Given:-

$$z = \frac{1}{16}$$

Equation 6.21

The Möbius estimate of e, eStack17, is then readily calculated from:-

$$e^z = \prod_{k=1}^{n} (1 - z^k)^{\frac{-\mu_k}{k}}$$

Equation 6.22

from which:-

$$e \approx \left[\prod_{k=1}^{n} (1 - z^k)^{\frac{-\mu_k}{k}} \right]^{\frac{1}{z}}$$

Equation 6.23

PSD(e_{fido},eStack17) is -0.00000000000018

A Few Classical Convergent Summations for e[6.7,6.8,6.9]

Sum of Reciprocal Factorials

The most basic series for Euler's Number, e, is the Sum of Reciprocal Factorials, ewiki1, given by:-

$$ewiki1 = \sum_{k=0}^{m2} \frac{1}{k!}$$

Equation 6.24

Adequate m2 is 17.
PSD(e_{fido},ewiki1) is -0.000000000000016

Reciprocal of Sum of Alternating Reciprocals

This is defined by:-

$$ewiki2 = \frac{1}{e_{est}} = \sum_{k=0}^{m2} \frac{(-1)^k}{k!}$$

Equation 6.25

where e_{est} is an Estimate of the Euler Number.
PSD(e_{fido},ewiki2) is -0.000000000000016 and accordingly, for the additional expense, there is no advantage over the simple sum of reciprocals ewiki1.

Sum of Adjusted Reciprocals #1

This conformation, ewiki4, requires m3 = 18, and is given by:-

$$ewiki4 = \frac{1}{2} \cdot \sum_{k=0}^{m3} \frac{(k+1)}{k!}$$

Equation 6.26

PSD(e_{fido},ewiki4) is zero.

Sum of Adjusted Reciprocals #2

This conformation, ewiki6, requires just half the iterations of ewiki4, so that m3 = 9, and is given by:-

$$ewiki6 = \sum_{k=0}^{m3} \frac{(3 - 4k^2)}{(2k + 1)!} = 2.71828182845905$$

Equation 6.27

PSD(e_{fido},ewiki6) is zero.
Equation 6.27 is identical to Brothers Equation 10 in the Brothers (2004) paper[6.3]

<u>Series for Unitary General Exponent of e: x = 1</u>

Note that for x = 1 and m2 = 17:-

$$ewiki12 = e^1 = \frac{2 + x}{2 - x} + \sum_{k=2}^{m2} \frac{-x^{(k+1)}}{k!.(k - x).(k + 1 - x)}$$

Equation 6.28

which resolves to:-

$$ewiki12 = e^1 = 3 - \sum_{k=2}^{m2} \frac{1}{k!.(k - x).(k + 1 - x)}$$

Equation 6.29

For both of the Equations 6.28 and 6.29 the PSD(e_{fido}, ewiki12) is - 0.000000121642391
In my opinion ewiki12 is inadequate.

<u>A Comparison of Selected Classical Convergent Series for e</u>

For eight of the classical methods discussed above I computed solution series according to the equation:-

$$\ln(|f'|) \approx \beta.k + \gamma$$

Equation 6.30

and in the case of ewiki6 also:-

$$\ln(|f'|) \approx \alpha k^2 + \beta.k + \gamma$$
Equation 6.12

All series appeared to be quadratic but not to the extent that rough linear forms were untenable. For some of the methods it was realistic to retain the linear model used with the first differentials of the convergent involving π or Ψ.

ewiki6 gave by far the most rapid and consistent convergence, whilst ewiki13 (Pippenger Product) and eStack17 (Möbius: x = 1/16) failed to conform to either Equation 6.30 or 6.12 despite (or because of?) strong convergence.

These eight methods are depicted in the plot Figure 6.2 and summary statistics in Table 6.3

For your convenience, but for illustrative purposes only, the requisite Iterations, k, for "acceptable" e in the best eight of the methods are:-

		k
ewiki13	Pippenger Product	7
eStack17	Möbius: x = 1/16	8 (12)
ewiki6	Sum of Adjusted Reciprocals #2	10
ewiki12	General Exponent of ex: x = 1	16
eStack16	Möbius: Direct Product	16
ewiki1	Sum of Reciprocal Factorials	17
ewiki2	Reciprocal of Sum of Alternating Reciprocals	18
ewiki4	Sum of Adjusted Reciprocals #1	19

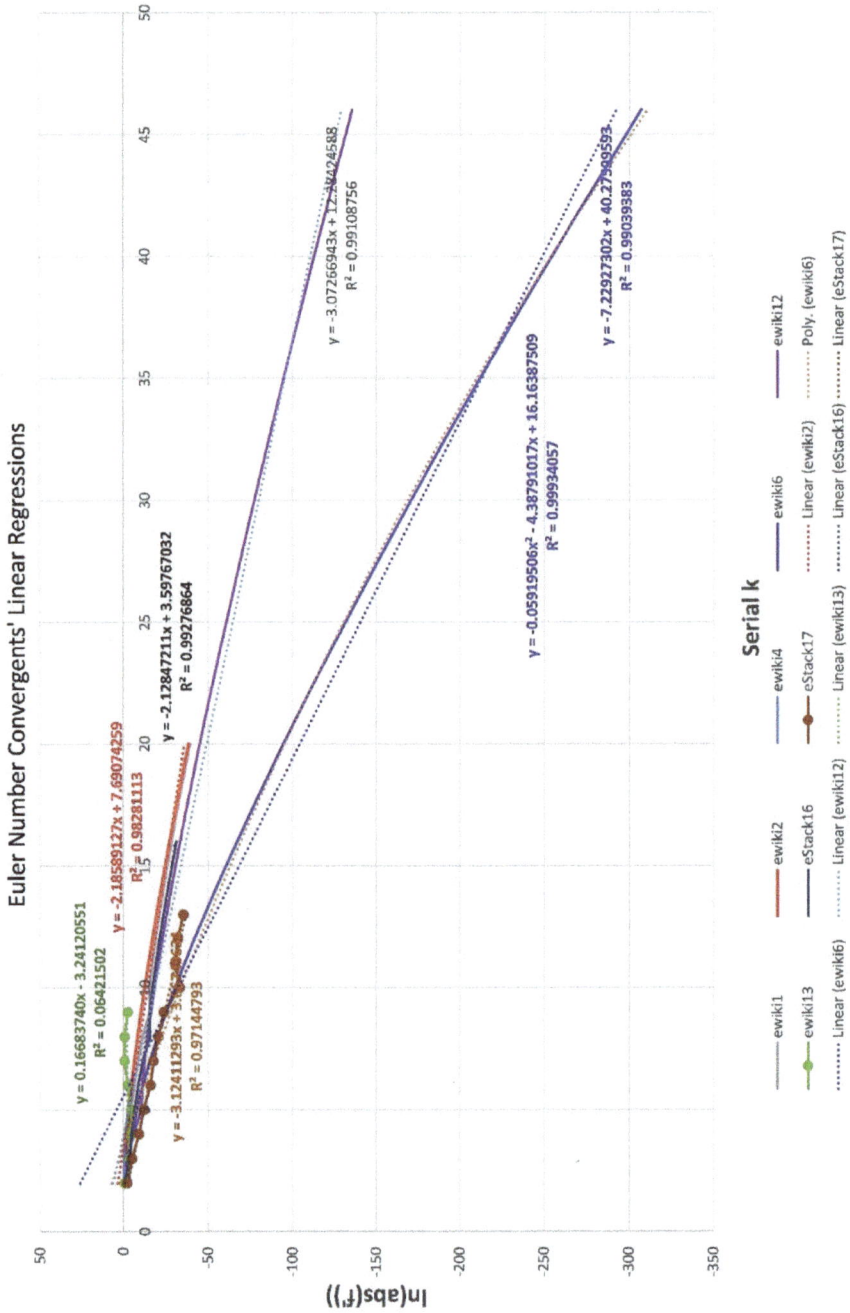

Euler Number Convergents' Linear Regressions

$y = 0.16683740x - 3.24120551$
$R^2 = 0.06421502$

$y = -2.18589127x + 7.69074259$
$R^2 = 0.9828113$

$y = -2.12847211x + 3.59767032$
$R^2 = 0.99276864$

$y = -3.07266943x + 12.74924588$
$R^2 = 0.99108756$

$y = -0.05919506x^2 - 4.38791017x + 16.16387509$
$R^2 = 0.99934057$

$y = -3.12411293x + 3.$
$R^2 = 0.97144793$

$y = -7.22927302x + 40.27940593$
$R^2 = 0.9903983$

ln(abs(f'))

Serial k

ewiki1	ewiki2
ewiki13	eStack16
Linear (ewiki6)	Linear (ewiki12)
ewiki4	ewiki6
eStack17	ewiki12
Linear (ewiki13)	Linear (eStack16)
Linear (ewiki2)	Poly. (ewiki6)
Linear (eStack16)	Linear (eStack17)

Figure 6.2
Classical Convergent Algorithms for e
ln(|f'|) versus Number of Iterations, k

Fiducial Values

The Ludolphine Constant	π	3.1415926535 8979
The Pythagoras Constant	ψ	1.4142135623 7310
The (Major) Ratio of Phidias	Φ	1.6180339887 4989
The Euler Constant	e	2.7182818284 5905

CONVERGENTS

	Monotonic Series	Alternating Series	Halved Series	Nine-Term Series	General Exponent of e^x Series: x = 1 (simplified)	Pippenger's Product	Gaillera-Sandow Theorem 5.3	Mobius Type Product: Direct Product	Mobius Type Product: Mobius Function: x = 1/16
	ewiki1	ewiki2	ewiki4	ewiki6	ewiki12	ewiki13	ewiki15	eStack16	eStack17
Sum or Product	2.7182818284 59050	0.3678794411 71442	5.4365636590 18090	2.7182818284 59050	0.2817181715 40955	0.7884720935 66756	Not Computed	2.4356365691 8090	1.0649945689 17860
Corrector	1.0000000000 0000	1.0000000000 0000	0.5000000000 0000			1.2682757048 67260	Not Computed	2.7182818284 59050	1.0649945689 17860
Result	2.7182818284 59050	2.7182818284 59040	2.7182818284 59050	2.7182818284 59050	2.7182818284 59050	53.3427442252 47900	Not Computed	2.4356365691 8090	1.0649945689 17860
PSD(fido, result)	-0.0000000000 0000216	0.0000000000 0000033	0.0000000000 0000000	0.0000000000 0000000	0.0000000000 0000000	Not Computed	Not Computed	10.3638323514 32700	-0.0000000000 000021

Linear Regression Parameters for the Count k (x) versus ln(|Γ'|)(y)

	ewiki1	ewiki2	ewiki4	ewiki6	ewiki12	ewiki13	ewiki15	eStack16	eStack17
Intercept	5.9430849 74	7.6907425 95	16.8293241 2	40.2759959 3	12.2842458 8	-3.2412055 08	Not Computed	3.5976703 19	3.6674963 43
Grade	-2.1732752 26	-2.1858912 71	-2.9057440 2	-7.2292730 23	-3.0726694 33	0.1668374 03	Not Computed	-2.1284271 08	-3.1241129 31
RSQ	0.9912153 71	0.9828111 33	0.9877567 227	0.9903938 31	0.9910875 57	0.0642150 23	Not Computed	0.9997686 37	0.9714479 3

Table 6.3
Summary Results for e Selected Approximations

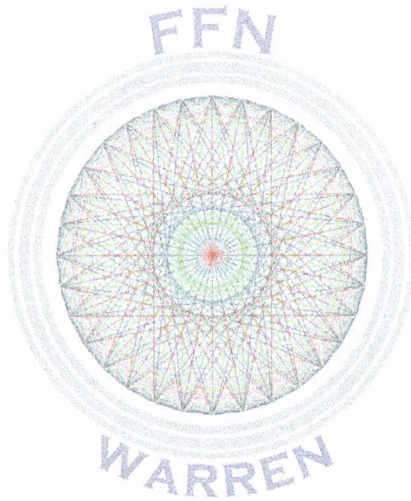

We shall define a Reciprocal Linear vulgar Fraction (RLF) as:-

$$RLF = \frac{n}{Mk + j}$$
Equation 7.1

where n, M and j are all Integer Constants, and k is an Integer Counter likely to vary.

In particular, we may examine pairs or triples of RLFs in which n and j vary according to:-

$$u = RLF_1 + RLF_2 + RLF_3 = \frac{n_1}{Mk + j_1} + \frac{n_2}{Mk + j_2} + \frac{n_3}{Mk + j_3}$$
Equation 7.2

Equation 7.2 is a triple-term object so in the case of a pair of fractions the expression $n_3/(Mk+j_3)$ is zero.

For demonstrative purposes allow that locally a = 23; b = 17; c = 34; d = 12; e = 63; f = 16; g = 33 and h = 359.

Employing this data in the addition of a pair of fractions we may state:-

$$u = \frac{1}{a} + \frac{1}{b} = \frac{b}{ab} + \frac{a}{ab} = \frac{b + a}{ab} = 0.10230179028133$$
Equation 7.3a

and:-

$$v = \frac{1}{d} + \frac{1}{e} = \frac{e}{de} + \frac{d}{de} = \frac{e + d}{de} = 0.099206349206349$$
Equation 7.3b

and for a triple:-

$$U = \frac{1}{a} + \frac{1}{b} + \frac{1}{c} = \frac{bc + ac + ab}{abc} = 0.131713554987212$$
Equation 7.4a

and:-

$$V = \frac{1}{d} + \frac{1}{e} + \frac{1}{f} = \frac{ef + df + de}{def} = 0.161706349206349$$
Equation 7.4b

The Addition of Fractions with Numerators

For the sake of demonstration further allow that locally:-

$$k = 1; \quad ii = 1; \quad jj = 4; \quad M = 8; \quad a = Mk + ii \text{ and } b = Mk + jj$$

Consequently:-

$$a = Mk + ii = 57$$
Equation 7.5a

and:-

$$b = Mk + jj = 60$$
Equation 7.5b

Biterm Case: Both Numerators Positive

Introduce the Numerators $n_1 = 4$ and $n_2 = 3$.
Then:-

$$u = \frac{n_1}{Mk + ii} + \frac{n_2}{Mk + jj} = \frac{4}{8k + 1} + \frac{3}{8k + 4} = \frac{n_1}{a} + \frac{n_2}{b}$$
$$= \frac{b.n_1}{ab} + \frac{a.n_2}{ab} = \frac{b.n_1 + a.n_2}{ab}$$
$$= 0.120175438596491$$
Equation 7.6

Biterm Case: One Numerator Negative

Introduce the Numerators $n_1 = 4$ and $n_2 = -2$.
Then:-

$$u = \frac{n_1}{Mk + ii} + \frac{n_2}{Mk + jj} = \frac{4}{8k + 1} - \frac{2}{8k + 4} = \frac{n_1}{a} + \frac{n_2}{b}$$
$$= \frac{b.n_1}{ab} + \frac{a.n_2}{ab} = \frac{b.n_1 + a.n_2}{ab}$$
$$= 0.036842105263158$$

Equation 7.7

Biterm Case: Both Numerators Negative

Introduce the Numerators $n_1 = -4$ and $n_2 = -7$.
Then:-

$$u = \frac{n_1}{Mk + ii} + \frac{n_2}{Mk + jj} = \frac{-4}{8k + 1} - \frac{7}{8k + 4} = \frac{n_1}{a} + \frac{n_2}{b}$$
$$= \frac{b.n_1}{ab} + \frac{a.n_2}{ab} = \frac{b.n_1 + a.n_2}{ab}$$
$$= -0.186842105263158$$

Equation 7.8

Triterm Case:- Three Numerators Positive

For example,

$$k = 8 \quad ii = 1 \quad jj = 4 \quad kk = 5 \quad M = 8$$
$$a = Mk + ii \quad b = Mk + jj \quad c = Mk + kk$$
$$n_1 = 4 \quad\quad n_2 = 7 \quad\quad n_3 = 91$$

Then:-

$$U = \frac{n_1}{a} + \frac{n_2}{b} + \frac{n_3}{c} = \frac{n_1}{MK + ii} + \frac{n_2}{Mk + jj} + \frac{n_3}{Mk + kk}$$
$$= \frac{n_1}{8 \times 8 + 1} + \frac{n_2}{8 \times 8 + 4} + \frac{n_3}{8 \times 8 + 5}$$
$$= \frac{b.c.n_1 + a.c.n_2 + a.b..n_3}{abc}$$
$$= 1.48332021771919$$

Equation 7.9

Triterm Case:- One Numerator Negative

For example,

$$n_1 = 4 \qquad n_2 = -7 \qquad n_3 = 91$$

Then:-

$$
\begin{aligned}
U = \frac{n_1}{a} + \frac{n_2}{b} + \frac{n_3}{c} &= \frac{n_1}{MK + ii} + \frac{n_2}{Mk + jj} + \frac{n_3}{Mk + kk} \\
&= \frac{n_1}{8 \times 8 + 1} + \frac{n_2}{8 \times 8 + 4} + \frac{n_3}{8 \times 8 + 5} \\
&= \frac{b.c.n_1 + a.c.n_2 + a.b..n_3}{abc} \\
&= 1.27743786477802
\end{aligned}
$$

Equation 7.10

Triterm Case:- Two Numerators Negative

For example,

$$n_1 = 4 \qquad n_2 = -7 \qquad n_3 = -91$$

Then:-

$$
\begin{aligned}
U = \frac{n_1}{a} + \frac{n_2}{b} + \frac{n_3}{c} &= \frac{n_1}{MK + ii} + \frac{n_2}{Mk + jj} + \frac{n_3}{Mk + kk} \\
&= \frac{n_1}{8 \times 8 + 1} + \frac{n_2}{8 \times 8 + 4} + \frac{n_3}{8 \times 8 + 5} \\
&= \frac{b.c.n_1 + a.c.n_2 + a.b..n_3}{abc} \\
&= -1.36024329464227
\end{aligned}
$$

Equation 7.11

Triterm Case:- Three Numerators Negative

For example,

$$n_1 = -4 \qquad n_2 = -7 \qquad n_3 = -91$$

Then:-

$$U = \frac{n_1}{a} + \frac{n_2}{b} + \frac{n_3}{c} = \frac{n_1}{MK + ii} + \frac{n_2}{Mk + jj} + \frac{n_3}{Mk + kk}$$

$$= \frac{n_1}{8 \times 8 + 1} + \frac{n_2}{8 \times 8 + 4} + \frac{n_3}{8 \times 8 + 5}$$

$$= \frac{b.c.n_1 + a.c.n_2 + a.b..n_3}{abc}$$

$$= -1.48332021771919$$

Equation 7.12

Additions of Fractions involving Split Denominators

In the case of a QuadTerm fraction addition we may write:-

$$X = \frac{a}{e} + \frac{b}{f} + \frac{c}{g} + \frac{d}{h} = \frac{afgh + begh + cefh + defg}{efgh}$$

Equation 7.13

Such a formation is exampled by the quadterm of the Bailey-Borwein-Plouffe formula for π:–

$$\frac{1}{16^k} BBPquadterm = \frac{1}{16^k}\left(\frac{4}{8k + 1} - \frac{2}{8k + 1} - \frac{1}{8k + 1} - \frac{1}{8k + 1}\right)$$

Equation 7.14

The multiplier $1/(16^k)$ is segregated for economy.
The full Bailey-Borwein-Plouffe[7.1,7.2] estimator for π is given by:-

$$\pi = \sum_{k=0}^{\infty} \frac{1}{16^k} BBPquadterm$$

$$= \sum_{k=0}^{\infty} \frac{1}{16^k}\left(\frac{4}{8k + 1} - \frac{2}{8k + 1} - \frac{1}{8k + 1} - \frac{1}{8k + 1}\right)$$

Equation 7.15

In practice the infinite limit of k can be set to ten for "adequate" resolution of π to around fifteen figures. The value of π that BBP yielded using m3 = 10 on MathCad® was 3.14159265358979

Quadterm Operations for the BBP π Formula

Let:-

M = 8	k = 5		
$j_1 = 1$	$j_2 = 4$	$j_3 = 5$	$j_4 = 6$
$n_1 = 4$	$n_2 = -2$	$n_3 = -1$	$n_4 = -1$
a = 4	b = -2	c = -1	d = -1
e = 8k+1	f = 8k +4	g = 8k+5	h = 8k +6

This data is consistent with the k = Fifth Iteration of the BBP solution for π.

Accordingly:-

$$X = \frac{a}{e} + \frac{b}{f} + \frac{c}{g} + \frac{d}{h} = \frac{afgh + begh + cefh + defg}{efgh} = Y$$
$$= 0.008145077498206$$

Equation 7.13

or:-

$$X = Y = \frac{Numer}{Denom}$$

Equation 7.16

where:-

$$Numer = afgh + begh + cefh + defg = 30416$$

Equation 7.17

and:-

$$Denom = efgh = 3734280$$

Equation 7.18

First Numerator Triple

$$fgh = (8k + 4)(8k + 5)(8k + 6)$$
$$= 8(64k^3 + 120k^2 + 74k + 15)$$
$$= 8(2^6 k^3 + 2^3.3.5. k^2 + 2.37. k + 3.5)$$
$$= 91080$$

Equation 7.19a

Define the Polynomial p_1 as:-

$$p_1 = 2^6 k^3 + 2^3.3.5.k^2 + 2.37.k + 3.5 = 11385$$
Equation 7.19b

Introducing $n_1 = 4$:-

$$n_1 f g h = M.n_1.p_1$$
$$= 8.4.(2^6 k^3 + 2^3.3.5.k^2 + 2.37.k + 3.5)$$
$$= 364320$$
Equation 7.19c

Second Numerator Triple

$$egh = (8k + 1)(8k + 5)(8k + 6)$$
$$= 8(64k^3 + 96k^2 + 41k + \frac{15}{4})$$
$$= 8(2^6 k^3 + 2^5.3.k^2 + 41.k + \frac{15}{4})$$
$$= 84870$$
Equation 7.20a

Define the Polynomial p_2 as:-

$$p_2 = 2^6 k^3 + 2^5.3.k^2 + 41.k + \frac{15}{4} = 10608.75$$
Equation 7.20b

Introducing $n_2 = -2$:-

$$n_1 egh = M.n_1.p_2$$
$$= 8.-2.(2^6 k^3 + 2^5.3.k^2 + 41.k + \frac{15}{4})$$
$$= -169740$$
Equation 7.20c

Third Numerator Triple

$$efh = (8k + 1)(8k + 4)(8k + 6)$$
$$= 8(64k^3 + 88k^2 + 34k + 3)$$
$$= 8(2^6k^3 + 2^3.11k^2 + 2.17.k + 3)$$
$$= 82984$$
Equation 7.21a

Define the Polynomial p3 as:-

$$p_3 = 2^6k^3 + 2^3.11.k^2 + 2.17.k + 3 = 10373$$
Equation 7.21b

Introducing n3 = -1:-

$$n_1fgh = M.n_1.p_3$$
$$= 8.-1.(2^6k^3 + 2^3.3.5.k^2 + 2.37.k + 3.5)$$
$$= -82984$$
Equation 7.21c

Fourth Numerator Triple

$$efg = (8k + 1)(8k + 4)(8k + 5)$$
$$= 8(64k^3 + 80k^2 + 29k + \frac{5}{2})$$
$$= 8(2^6k^3 + 2^4.5.k^2 + 29.k + \frac{5}{2})$$
$$= 81180$$
Equation 7.22a

Define the Polynomial p4 as:-
$$p_4 = 2^6k^3 + 2^4.5.k^2 + 29.k + \frac{5}{2} = 10147.5$$
Equation 7.22b

Introducing n1 = -2:-

$$n_1fgh = M.n_1.p_4$$
$$= 8.-2.(2^6k^3 + 2^4.5.k^2 + 29.k + \frac{5}{2})$$
$$= -81180$$
Equation 7.22c

The Numerator and the Denominator of the Iterate

By substitution in Equation 7.17:-

$$Numer = n_1 fgh + n_2 egh + n_3 efh + n_4 efg = 30416$$

Equation 7.23

or:-

$$Numer = Mn_1 p_1 + Mn_2 p_2 + Mn_3 p_3 + Mn_4 p_4$$
$$= M \sum_{i=1}^{4} n_i p_i$$

Equation 7.24

Meanwhile:-

$$Denom = (Mk + j_1)(Mk + j_2)(Mk + j_3)(Mk + j_4)$$
$$= \sum_{i=0}^{4} t_i$$

Equation 7.25

where:-

$$t_0 = \prod_{i=1}^{4} j_i = 120$$

Equation 7.26a

$$t_1 = (Mk) \sum_{m=1}^{4} \frac{\prod_{l=1}^{4} j_l}{j_m} = 7760$$

Equation 7.26b

Because:-

$$v = j_1 j_2 + j_1 j_3 + j_1 j_4 + j_2 j_3 + j_2 j_4 + j_3 j_4$$
$$= j_1 \sum_{l=2}^{4} j_l + j_2 \sum_{l=3}^{4} j_l + j_3 \sum_{l=4}^{4} j_l$$

$$= \sum_{l=1}^{3} j_l \sum_{m=l+1}^{4} j_m = 89$$
$$\textbf{Equation 7.27}$$

We obtain:-

$$t_2 = (Mk)^2 \sum_{l=1}^{3} j_l \sum_{m=l+1}^{4} j_m = 142400$$
$$\textbf{Equation 7.26c}$$

$$t_3 = (Mk)^3 \sum_{k=1}^{4} j_k = 1024000$$
$$\textbf{Equation 7.26d}$$

$$t_4 = (Mk)^4 = 2560000$$
$$\textbf{Equation 7.26e}$$

Thus by substitution:-

$$Denom = (MK)^0 . \prod_{i=1}^{4} j_i + (MK)^1 . \sum_{m=1}^{4} \frac{\prod_{l=1}^{4} j_l}{j_m}$$
$$+ (Mk)^2 \sum_{l=1}^{3} j_l \sum_{m=l+1}^{4} j_m + (MK)^3 . \sum_{k=1}^{4} j_k + (MK)^4$$
$$\textbf{Equation 7.28}$$

Equation 7.28, the Denominator of the Iterate, is a fourth-degree algebraic polynomial in (Mk), whilst previous developments confirmed that Numer is a third-degree (cubic) polynomial.

Therefore, the general equation for a Constant is satisfied as:-

$$C = \sum_{k=0}^{\infty} K_k = \sum_{k=0}^{\infty} \frac{1}{16^k} . Y$$
$$\textbf{Equation 7.29}$$

so:-

$$C = \sum_{k=0}^{\infty} \frac{1}{16^k} \cdot \frac{afgh + begh + cefh + defg}{efgh}$$

$$= \sum_{k=0}^{\infty} \frac{1}{16^k} \cdot \left(\frac{4}{8k+1} - \frac{2}{8k+1} - \frac{1}{8k+1} - \frac{1}{8k+1} \right)$$

$$= \sum_{k=0}^{\infty} \frac{1}{16^k} \cdot \frac{Numer_k}{Denom_k}$$

$$= \pi$$

Equation 7.30

or:-

$$C = \sum_{k=0}^{\infty} \frac{1}{b^k} \cdot \frac{P(k)}{Q(k)}$$

Equation 7.31

where C is some Constant; K_k is BBPquadterm, the Unadjusted Dividend of two Algebraic Polynomials; P(k) and Q(k) are Algebraic Polynomials; k is a counter; and b is the "Base", expressive of the digit-wise expression of a BBP-type formula (not operative here, where we work in denary).

Where the test k = 5, the local value of K_k is 0.000000007767751 and the local value of Y is 0.008145077498206

As previously remarked, the "infinite" limit of the major summation is not truly "infinite" in engineering real-life: About ten iterations will give you an "adequate" estimate good to thirteen or fifteen figures on a 64-bit system.

Efficacy of The Dividend Numer/Denom

By substitution 7.23 (Numer) and 7.25 (Denom) in the k-th element of Equation 7.29 we obtain a large formula.

This large formula is shown in Figure 7.1 as Equation 7.32 for clarity. A more succinct re-arrangement is presented as Equation 7.33 in Figure 7.2

At this juncture it is helpful to define:-

$$sj = \sum_{k=1}^{4} j_k = 16$$

Equation 7.34a

$$pj = \prod_{k=1}^{4} j_k = 120$$

Equation 7.34b

$$rj = \sum_{k=1}^{4} \frac{1}{j_k} = 1.61666666666667$$

Equation 7.34c

$$N = Mk = 40$$

Equation 7.34d

From these it follows that Equation 7.33 may be simplified by substitution to Equation 7.35 as presented in Figure 7.3

Alternatives for the $(M.k)^2$ Multiplier

Let J be the multiplier of $(Mk)^2$ in Equation 7.32 Furthermore, allow that:-

$v_1 = j_1 j_2 = 4$		**Equation 7.36a**
$v_2 = j_1 j_2 = 5$		**Equation 7.36b**
$v_3 = j_1 j_4 = 6$		**Equation 7.36c**
$v_4 = j_2 j_3 = 20$		**Equation 7.36d**
$v_5 = j_2 j_4 = 24$		**Equation 7.36e**
$v_6 = j_3 j_4 = 30$		**Equation 7.36f**

Then:-

$$J = \sum_{l=1}^{3} j_l \sum_{m=l+1}^{4} j_m = \sum_{ii=1}^{3} \left(j_{ii} \sum_{jj=ii+1}^{4} j_{jj} \right) = \left(\sum_{l=1}^{4} j_i \right) \left(\sum_{m=1}^{6} \frac{1}{v_m} \right)$$

Equation 7.37

$$K_k := \frac{1}{16^k} \cdot \frac{M \cdot \sum_{i=1}^{4} n_i \cdot p_i}{(M \cdot k)^0 \cdot \prod_{i=1}^{4} j_i + (M \cdot k) \cdot \sum_{m=1}^{4} \frac{\prod_{l=1}^{4} j_l}{j_m} + (M \cdot k)^2 \cdot \sum_{l=1}^{3} j_l \cdot \sum_{m=l+1}^{4} j_m + (M \cdot k)^3 \cdot \sum_{k=1}^{4} j_k + (M \cdot k)^4}$$

Figure 7.1
Equation 7.32

$$K_k := \frac{1}{16^k} \cdot \frac{\prod_{i=1}^{4} j_i \cdot \left((M \cdot k)^0 + (M \cdot k) \cdot \sum_{m=1}^{4} \frac{1}{j_m} \right) + (M \cdot k)^2 \cdot \left(\sum_{l=1}^{3} j_l \cdot \sum_{m=l+1}^{4} j_m + (M \cdot k) \cdot \sum_{k=1}^{4} j_k + (M \cdot k)^2 \right)}{M \cdot \sum_{i=1}^{4} n_i \cdot p_i}$$

Figure 7.2
Equation 7.33

$$K_k := \frac{1}{16^k} \cdot \frac{M \cdot \sum_{i=1}^{4} n_i \cdot p_i}{pj \cdot (1 + N \cdot rj) + N^2 \cdot \left(\sum_{l=1}^{3} j_l \cdot \sum_{m=l+1}^{4} j_m + N \cdot sj + N^2\right)}$$

Figure 7.3
Equation 7.35

On the face of it the right-most of the alternatives for J is the least computationally-efficient but if meantime we adopt it for substitution we may write:-

$$K_k = \frac{1}{16^k} \times \frac{M \sum_{i=1}^4 n_i p_i}{pj.(1+N.j) + N^2\left[\left(\sum_{l=1}^4 j_i\right)\left(\sum_{m=1}^6 \frac{1}{v_m}\right) + N.sj + N^2\right]}$$

Equation 7.38

which reduces to:-

$$K_k = \frac{1}{16^k} \times \frac{M \sum_{i=1}^4 n_i p_i}{pj.\left[1 + N.rj + N^2.\left(\sum_{m=1}^6 \frac{1}{v_m}\right)\right] + N^2[N.sj + N^2]}$$

Equation 7.39

or:-

$$K_k = \frac{1}{16^k} \times \frac{M \sum_{i=1}^4 n_i p_i}{pj.\left[1 + N.rj + N^2.\left(\sum_{m=1}^6 \frac{1}{v_m}\right)\right] + N^3.sj + N^4}$$

Equation 7.40

Back-substitution allows us to specify:-

$$K_k = \frac{1}{16^k}$$
$$\times \frac{M \sum_{i=1}^4 n_i p_i}{\prod_{k=1}^4 j_k.\left[1 + N.\sum_{k=1}^4 \frac{1}{j_k} + N^2.\left(\sum_{m=1}^6 \frac{1}{v_m}\right)\right] + N^3.\sum_{k=1}^4 j_k + N^4}$$

Equation 7.41

CHAPTER EIGHT
POLYNOMIAL INTEGRATION AND DIVISION
IN TERMS OF
PARTIAL FRACTIONS

In their classic paper "On the rapid computation of various polylogarithmic constants."[7.2] Bailey, Borwein and Plouffe discuss the general computation of Constants in terms of:-

$$CC = \sum_{k=1}^{\infty} \frac{p(k)}{\beta^{ck} \cdot q(k)}$$
Equation 8.1

and in particular they present:-

$$\pi = \sum_{i=0}^{\infty} \frac{1}{16^i} \left(\frac{4}{8i+1} - \frac{2}{8i+4} - \frac{1}{8i+5} - \frac{1}{8i+6} \right)$$
Equation 8.2

CC is some real Constant; β is the "Base", which can be employed in the selection of particular digits from a number mantissa (not applied by us: We work exclusively in denary complete values); c and k are both integers, k being the Iteration Counter; and p(k) and q(k) are Algebraic Polynomials in Integer k. (I.e. p(k) and q(k) are Diophantine Equations).

q(k) must be of higher degree than p(k).

π is the Ludolphine Constant.

In our idiom we may at the expense of a slight redundancy conveniently re-generalise Equation 8.1 as:-

$$CC = \frac{1}{r_{con}} \sum_{k=0}^{\infty} \left(\frac{1}{\beta^k} \times \sum_{j=0}^{J} \frac{n_j}{M.j + d_j} \right)$$
Equation 8.3

Given that:-

M = 8	J = 4	β = 16	r_{con} = 1
n_1 = 4	n_2 = -2	n_3 = -1	n_4 = -1
d_1 = 1	d_2 = 4	d_3 = 5	d_4 = 6

Equation 8.2 becomes:-

$$\pi = \frac{1}{1}\sum_{k=0}^{\infty}\left(\frac{1}{16^k} \times \sum_{j=0}^{J}\frac{n_j}{8.j + d_j}\right)$$

$$= \sum_{k=0}^{\infty}\frac{1}{16^k}\left(\frac{4}{8k+1} - \frac{2}{8k+4} - \frac{1}{8k+5} - \frac{1}{8k+6}\right)$$

Equation 8.4

In the subsequent development:-

$A \equiv n_1$ $\qquad$ $B \equiv n_2$ $\qquad$ $C \equiv n_3$ $\qquad$ $D \equiv n_4$

whilst:-

$a \equiv d_1$ $\qquad$ $b \equiv d_2$ $\qquad$ $c \equiv d_3$ $\qquad$ $d \equiv d_4$

x should normally be assumed unity for analytic and numerical purposes.

Some Procedural Hazards

(1) **"Infinity" is Fortunately a Small Count Limit**
As a rule, ten is the largest iteration limit you will need to achieve thirteen or more digits on a 64-bit system, at least with Equation 8.1 and its BBP-type progeny.
(2) **Division by Zero**
Equation 8.1 and its BBP-type special equations are capable of term-wise "Division by Zero" (theoretically an infinite quotient).
This rarely bothers theoreticians but it certainly frustrates practical computerists.
For example, in a BBP-type formula in which M = 6, and the kth. iteration sees k = 2, and if d_2 is -12, then (M.k-d_2) would be zero and computation would fail as "division by zero".
(3) **The Quotient of Two Algebraic Polynomials is *Not* a Polynomial**
This does not mean that a numerical quotient is unavailable and as we shall see the art of Partial Fractions can be used to transform Equation 8.1 BBP-type formulae to consolidated

coefficient-driven algebraic polynomials susceptible to analysis and processing as such.

Also, the fact that the quotient is not a polynomial renders Newton-Cotes formulae and other numerical integration methods that *assume* that the subject is an algebraic polynomial inaccurate. But not necessarily inaccurate enough to be misleading, because numerical integrations are themselves approximants, and, with due latitude, can be employed to "confirm" symbolic integrals, etcetera.

(4)   **M.k is Too Big**

During some phases of analysis and development "x" needs to be treated as unity rather than as M.k

As with all things in mathematics and science one must use all of your skill and discretion or else your research will turn to dust in your hands.

On the other hand, eventual world renown (or whatever appeals) awaits success, or even brilliant failure, for we remember our Empedocles' as well as our Einsteins.

Transformation of the BBP Formula for Pi

The quotient of two univariate polynomials, locally called FF and GG, can be transformed into a Rational Fraction $RR(x)^{8.2}$:-

$$RR(x) = \frac{FF}{GG}$$

Equation 8.5

It is now handy to quote Equation 8.4 as:-

$$F_k = \sum_{k=0}^{m3} \frac{1}{16^k}\left(\frac{A}{x+a} - \frac{B}{x+b} - \frac{C}{x+c} - \frac{D}{x+d}\right)$$

Equation 8.6

where x = M.k and A, B, C and D are Integer Numerators of their respective quotients. It is only necessary that m3 is 10 rather than infinity and the Equation 8.6 can be transformed. (Numerator Constants B, C, and D are negative for the BBP π formula).

For example, at the iteration where k = 5, M×k is *at that point* 8×5 = 40 in the context of the BBP formula for π, and F_k may be rendered as:-

$$F_5 = \sum_{k=5}^{5} \left[\frac{1}{\beta^k} \cdot \frac{p(k)}{q(k)} \right]$$

$$= \frac{1}{\beta^k} \cdot \frac{120k^2 + 151k^1 + 47k^0}{512k^4 + 1024k^3 + 712k^2 + 194k^1 + 15k^0}$$

Equation 8.7

F_5 has the numerical value 0.000000007767751, as is confirmed by the rubricated figure in Table 8.1 which presents my EXCEL® elaboration of the cumulated BBP-type solution for π.

The ways in which p(k)/q(k) is established will be discussed below.

Now:-

$$F_5 = \sum_{k=5}^{5} \left[\frac{1}{\beta^k} \cdot \frac{p(k)}{q(k)} \right] = \frac{1}{\beta^k} \cdot \frac{\sum_{i=1}^{3} c_i . k^i}{\sum_{i=1}^{5} \kappa_i . k^i}$$

Equation 8.8

where c_i and κ_i are Integer Coefficients to be determined.

We may for convenience exclude the k-dependent multiplier $1/\beta^k$ from aspects of our analysis when we need to focus upon the four-quotient core or "kernel" of F_k, a group of fractions we may designate G_k, (given -ve B, C, D):-

$$G_k = \frac{A}{x+a} + \frac{B}{x+b} + \frac{C}{x+c} + \frac{D}{x+d}$$

$$= \frac{A}{M.k+a} + \frac{B}{M.K+b} + \frac{C}{M.k+c} + \frac{D}{M.k+d}$$

$$= \frac{A}{8k+1} + \frac{B}{8k+4} + \frac{C}{8k+5} + \frac{D}{8k+6}$$

Equation 8.9

such that:-

$$F_k = \frac{G_k}{16^k} = \frac{1}{16^k} \times \frac{A.s + B.t + C.u + D.v}{r}$$
$$= 0.000000007767751$$

Equation 8.10

where s, t, u, v and r are all Diophantine Polynomials to be determined.

The fifth iteration is:-

$$G_5 = \frac{4}{8k+1} - \frac{2}{8k+4} - \frac{1}{8k+5} - \frac{1}{8k+6}$$
$$= \frac{4}{8 \times 5 + 1} - \frac{2}{8 \times 5 + 4} - \frac{1}{8 \times 5 + 5} - \frac{1}{8 \times 5 + 6}$$
$$= \frac{4}{41} - \frac{2}{44} - \frac{1}{45} - \frac{1}{46}$$
$$= 0.008145077498206$$

Equation 8.11

This value is shown in blue in Table 8.1

Establishment of the Consolidated Diophantine Polynomials

These are the consolidated denominator r of Equation 8.10, and the four numerator constants s, t, u, and v.

At this juncture it is wise to replace M.k with x = unity until we are in a position to restore the proper values.

We may use cross-multiplication and cancellation to render G_k in the following common-denominator form as the large Equation 8.12 represented by Figure 8.1

The bracketed sums can be multiplied and simplified as follows:-

Denominator

r = (x-a)(x-b)(x-c)(x-d)
$r_1 = x^4 + x^3(a+b+c+d) + x^2(ab+ac+ad+bc+bd+cd)$
$r_2 = x(abc+abd+acd+bcd) + abcd$
$r = r_1 + r_2$
$r = x^4 + 16x^3 + 89x^2 + 194x + 120 = 3734280$

FIDUCIAL PI: 3.14159265358979 0
CUMULATE PI: 3.14159265358979 0
PSD: 0.00000000000000

		4	-2	-1	-1
		8k+1	8k+4	8k+5	8k+6
Numerator n		4	-2	-1	-1
Multipicand M		8	8	8	8
Augend j		1	4	5	6

k	$1/16^k$	Cumulate Sum	$(1/16k)*$ Sum of Fractions	Sum of Fractions	8k+1	8k+4	8k+5	8k+6
0	1.00000000000000	3.13333333333330	3.13333333333330	3.13333333333330	4.000000	-0.500000	-0.200000	-0.166667
1	0.06250000000000	3.14142246642470	0.00808091330891 33	0.12942612942612 9	0.444444	-0.166667	-0.076923	-0.071429
2	0.00390625000000	3.14158739034658 0	0.00016492392411 5	0.04222052457346 6	0.235294	-0.100000	-0.047619	-0.045455
3	0.00024414062500	3.14159245757440	0.00000506722085 4	0.02075533661740 6	0.160000	-0.071429	-0.034483	-0.033333
4	0.00001525878906 3	3.14159264546340	0.00000018789290 1	0.01213137491558 54	0.121212	-0.055556	-0.027027	-0.026316
5	0.00000095367431 6	3.14159265322809 0	0.00000000076775 1	0.00814507749820 6	0.097561	-0.045455	-0.022222	-0.021739
6	0.00000005960464 5	3.14159265357288 0	0.00000000003447 93	0.00578467155286 6	0.081633	-0.038462	-0.018868	-0.018519
7	0.00000000372529 0	3.14159265358897 50	0.00000000001609 2	0.00431963038214 3	0.070175	-0.033333	-0.016393	-0.016129
8	0.00000000023283 1	3.14159265358975 0	0.00000000000078 0	0.00334828923677	0.061538	-0.029412	-0.014493	-0.014286
9	0.00000000001455 2	3.14159265358979 0	0.00000000000003 9	0.00267120526673 5	0.054795	-0.026316	-0.012987	-0.012821
10	0.00000000000090 9	3.14159265358979 0	0.00000000000000 02	0.00218057938076 2	0.049383	-0.023810	-0.011765	-0.011628
11	0.00000000000005 7	3.14159265358979 0	0.00000000000000 00	0.00181370374555 3	0.044944	-0.021739	-0.010753	-0.010638
12	0.00000000000000 4	3.14159265358979 0	0.00000000000000 00	0.00153272017344 25	0.041237	-0.020000	-0.009901	-0.009804
13	0.00000000000000 0	3.14159265358979 0	0.00000000000000 00	0.00131149855920 5	0.038095	-0.018519	-0.009174	-0.009091
14	0.00000000000000 0	3.14159265358979 0	0.00000000000000 00	0.00113526599599 56	0.035398	-0.017241	-0.008547	-0.008475
15	0.00000000000000 0	3.14159265358979 0	0.00000000000000 00	0.00099923110450 97	0.033058	-0.016129	-0.008000	-0.007937
TOTALS:	1.06666666666667 0	50.25704748223840 0	3.14159265358979 0	3.37128344715481 0	5.568768	-1.184065	-0.529155	-0.484264

Table 8.1
The BBP-type Solution for Pi

Numerator Term A

$$s = (x+b)(x+c)(x+d)$$
$$s = x^3 + x^2(b+c+d) + x(bc +bd + cd) + bcd$$
$$b+c+d = 15 \qquad bc + bd + cd = 74 \qquad bcd = 120$$
$$s = x^3 + 15x^2 + 74x +120 = 91080$$
$$s/r = 0.024390243902439$$

Numerator Term B

$$t = (x+a)(x+c)(x+d)$$
$$t = x^3 + x^2(a+c+d) + x(ac +ad + cd) + acd$$
$$a+c+d = 12 \qquad ac + ad + cd = 41 \qquad acd = 30$$
$$t = x^3 + 12x^2 + 41x +30 = 84870$$
$$t/r = 0.022727272727273$$

Numerator Term C

$$u = (x+a)(x+b)(x+d)$$
$$u = x^3 + x^2(a+b+d) + x(ab +ad + bd) + abd$$
$$a+b+d = 11 \qquad ab + ad + bd = 34 \qquad abd = 24$$
$$u = x^3 + 11x^2 + 34x +24 = 82984$$
$$u/r = 0.022222222222222$$

Numerator Term D

$$v = (x+a)(x+b)(x+c)$$
$$v = x^3 + x^2(a+b+c) + x(ab +ac + bc) + abc$$
$$a+b+c = 10 \qquad ab + ac + bc = 29 \qquad abc = 20$$
$$v = x^3 + 10x^2 + 29x +20 = 81180$$
$$v/r = 0.021739130434783$$

It is now apparent that:-

$$G_5 = \frac{A}{8k+a} - \frac{B}{8k+b} - \frac{C}{8k+c} - \frac{D}{8k+d}$$

$$= \frac{4}{8 \times 5 + 1} - \frac{2}{8 \times 5 + 4} - \frac{1}{8 \times 5 + 5} - \frac{1}{8 \times 5 + 6}$$

$$= \frac{As}{t} - \frac{Bt}{t} - \frac{Cu}{t} - \frac{Dv}{t}$$

$$= 0.008145077498206$$

Equation 8.13

Figure 8.2 presents the fully substituted equation for G_5.

Now we happen to know the values of A, B, C, D, and a, b, c, d because Bailey, Borwein and Plouffe have told us, but we could find ourselves in a general situation where such constants were to be determined.

Accordingly we shall review the process for the determination of the numerator constants A, B, C and D, leaving the denominator constants to be the subjects of theory.

The determination of the Numerator Constants is undertaken by:-

(A) The gathering of numerator polynomial coefficients by degree

(B) The solution of the four resulting linear equations

$$G_k = \frac{A(x+b)(x+c)(x+d) + B(x+a)(x+c)(x+d) + C(x+a)(x+b)(x+d) + D(x+a)(x+b)(x+c)}{(x+a)(x+b)(x+c)(x+d)(x+a)(x+b)(x+c)(x+d)}$$

Figure 8.1
Equation 8.12

$$G_5 = \frac{A(x+b)(x+c)(x+d) + B(x+a)(x+c)(x+d) + C(x+a)(x+b)(x+d) + D(x+a)(x+b)(x+c)}{(x+a)(x+b)(x+c)(x+d)}$$

Figure 8.2
Equation 8.14

Gathering Coefficients by Degree

Study of the four numerators' polynomial expansions for s, t, u and v confirms that there are four lots of x^3 each with an implicit coefficient of unity *when x is one*. Counting across the polynomials for s, t, u and v:-

I.e.:-

$$x^3(1.A + 1.B + 1.C + 1.D) = 0$$
Equation 8.15a

and also there are 48 lots of x^2 shared amongst A, B, C and D as in Equation 8.15b:-

$$x^2(15.A + 12.B + 11.C + 10.D) = 15$$
Equation 8.15b

and for x^1 and x^0 the share-outs below:-

$$x^1(74.A + 41.B + 34.C + 29.D) = 151$$
Equation 8.15c

$$x^0(120.A + 30.B + 24.C + 20.D) = 376$$
Equation 8.15d

And we can re-arrange Equation 8.10 as:-

$$G_k.r = r.\frac{F_k}{16^k} = A.s + B.t + C.u + D.v = 30416$$
Equation 8.16

or having the known values of s, t, u and v to hand:-

$$G_k.r = r.\frac{F_k}{16^k}$$
$$= A.s + B.t + C.u + D.v$$
$$= 91080A + 84870B + 82984C + 81180D$$
$$= 30416$$
Equation 8.17

Solving for A, B, C and D Using Matrix Algebra

Linear Equations such as the four Equations 8.15 can be solved by a number of methods, all of which attempt to find an equal number of multiplier constants whose values render each of the four equations valid. Hence we seek the four constants A, B, C and D on the assumption we do not already know their values.

Choice of method depends upon convenience, cost, accuracy and the technicalities of numerical stability during processing.

We shall apply the matrix algebra intrinsics as supplied as part of MathCad® Express®. These are easily adequate to small tutorial demonstrations.

The essential principle is to "plug" the known coefficients of the four Equations 8.15 into a 4×4 element 2D matrix we can call AA; to invert AA to attain AA⁻¹; and then to multiply the matrix AA⁻¹ by XX, a single-column 4×1 matrix of the numerical values of the 8.15 equations. This vector XX contains four integers.

The result is the 4×1 matrix BB that reports the four solution coefficients in order.

In formal terms:-

$$AA = \begin{bmatrix} 1 & 1 & 1 & 1 \\ 15 & 12 & 11 & 10 \\ 74 & 41 & 34 & 29 \\ 120 & 30 & 24 & 20 \end{bmatrix}$$

Equation 8.18

$$AA^{-1} = \begin{bmatrix} -0.0167 & 0.0167 & -0.0167 & 0.0167 \\ 10.6667 & -2.6667 & 0.6667 & -0.1667 \\ -31.25 & 6.25 & -1.25 & 0.25 \\ 21.6 & -3.6 & 0.6 & -0.1 \end{bmatrix}$$

Equation 8.19

(Elements of Equation 8.19 are shown only to four-figure accuracy).

$$\det(AA) = 120$$
Equation 8.20

$$XX = \begin{bmatrix} A \\ B \\ C \\ D \end{bmatrix}$$

Equation 8.21

$$BB = \begin{bmatrix} 0 \\ 15 \\ 151 \\ 376 \end{bmatrix}$$

Equation 8.22

Proceeding to solution:-

$$CC = \det(AA) \cdot AA^{-1} = \begin{bmatrix} -2 & 2 & -2 & 2 \\ 1280 & -320 & 80 & -20 \\ -3750 & 750 & -150 & 30 \\ 2592 & -432 & 72 & -12 \end{bmatrix}$$

Equation 8.23

and:-

$$SS = AA^{-1} \cdot BB = \begin{bmatrix} 4 \\ -2 \\ -0.999999999999986 \\ -1.00000000000003 \end{bmatrix}$$

Equation 8.24

or:-

$$SS = \frac{1}{\det(AA)} \cdot CC \cdot BB = \begin{bmatrix} 4 \\ -2 \\ -0.999999999999986 \\ -1.00000000000003 \end{bmatrix}$$

Equation 8.25

The solution vector SS is polluted by numerical error and the cogent solution is:-

$$SS = AA^{-1} \cdot BB = \begin{bmatrix} 4 \\ -2 \\ -1 \\ -1 \end{bmatrix}$$

Equation 8.26

where:-

$$A = SS_{0,0} = 4$$
Equation 8.27a

$$B = SS_{1,0} = -2$$
Equation 8.27b

$$C = SS_{2,0} = -1$$
Equation 8.27c

$$D = SS_{3,0} = -1$$
Equation 8.27d

Using Express® as stated the prior and posterior PSD errors in F and G were 0.000000000007667 percent.

Having knowledge of all constants it is now possible to re-arrange Equation 8.10 for general G_k in terms of the quotient of two algebraic polynomials. *Note that this quotient is not of itself an algebraic polynomial.* So:-

$$G_k = \frac{F_k}{16^k} = \frac{120k^2 + 151k + 47}{512k^4 + 1024k^3 + 712k^2 + 194k + 15}$$
Equation 8.28

For the fifth iterate this is:-

$$G_5 = \frac{F_5}{16^5} = \frac{120k^2 + 151k + 47}{512k^4 + 1024k^3 + 712k^2 + 194k + 15}$$
Equation 8.29

where the RHS k is 5.
Numerically, G_5 is 0.008145077498206

Status when Iteration x = 5

We can now conveniently redefine as M.j = 8×5 = 40 since we no longer need to guard against corrupting matrix operations.

So for numerical validation purposes at the fifth iterate we will use x = k =5 as j is also five for the fifth iterate (or the sixth if you count k = 0).

To be explicit we are reverting to the former conditions when x = 5, a = 1, b = 4, c = 5, d = 6, β = 16, M = 8 and k = 5.

So in these terms:-

$$u = x - a = 4$$
Equation 8.30a

$$v = x - b = 1$$
Equation 8.30b

$$w = x - c = 0$$
Equation 8.30c

$$y = x - d = -1$$
Equation 8.30d

whilst:-

$$G_5 = \frac{A}{8x + a} + \frac{B}{8x + b} + \frac{C}{8x + c} + \frac{D}{8x + d} = 0.008145077498205$$
Equation 8.31

or equivalently:-

$$G_5 = \frac{120k^2 + 151k + 47}{512k^4 + 1024k^3 + 712k^2 + 194k + 15} = 0.008145077498205$$
Equation 8.32

The Integration of F

Furthermore, since ten is an adequate surrogate of "infinity" in m3 = 10, we may define:-

$$F_k = func(k) = \frac{1}{16^k} \times \frac{120k^2 + 151k + 47}{512k^4 + 1024k^3 + 712k^2 + 194k + 15}$$
Equation 8.33

from which:-

$$Ι\pi = \sum_{k=0}^{m3} func(k) = \pi$$
Equation 8.34

and:-

$$Ip = \int_{0}^{m3} func(k) = 0.3536568831168$$
Equation 8.35

Therefore:-

$$F = func(5) = 0.000000007767751$$
Equation 8.36

It follows that IC1 is:-

$$IC1 = \frac{Ι\pi}{Ip} = \frac{\sum_{k=0}^{m3} func(k)}{\int_{0}^{m3} func(k).dk} = 8.88316558666339$$
Equation 8.37

and:-

$$IC2 = \frac{15}{47}\pi = 1.00263595327334$$
Equation 8.38

Accordingly, if Iq is the Integral of F as given by Equation 8.35 then:-

$$Iq = \int_{0}^{\infty} \frac{1}{16^k} \cdot \left(\frac{A}{8k+a} - \frac{B}{8k+b} - \frac{C}{8k+c} - \frac{D}{8k+d} \right) . dk$$
$$= 0.353656830100015$$
Equation 8.39

This value of 0.353656830100015 is identical to that I established using Wolfram® Alpha® whilst the PSD of the Iq value relative to the integral of func(k) is given by:-

$$PSD\left(Iq, \int_0^{m3} func(k).dk\right) = -0.000014991025434$$

Equation 8.40

In Equation 8.39 it is "adequate" if the upper-bounding limit "infinity" is taken numerically to be 128.

It is also possible to express the Integral of F as the sums of four definite integrals:-

$$Ir = \frac{15}{47}\left(A\int_0^1 \frac{1}{a}.du + B\int_0^1 \frac{1}{b}.dv + C\int_0^1 \frac{1}{c}.dw + D\int_0^1 \frac{1}{d}.dy\right)$$
$$= 1$$

Equation 8.41

which has the identity:-

$$Ir2 = \frac{15}{47}\left(\frac{A}{a} + \frac{B}{b} + \frac{C}{c} + \frac{D}{d}\right) = 1$$

Equation 8.42

For the BBP formula for π this is numerically:-

$$Ir2 = \frac{15}{47}\left(\frac{4}{1} + \frac{-2}{4} + \frac{-1}{5} + \frac{-1}{6}\right) = \frac{15}{47}\left(\frac{47}{15}\right) = 1$$

Equation 8.43

To express this scaled sum as π we can trivially multiply both sides by π:-

$$Ir2 = \frac{15}{47}.\pi.\left(\frac{4}{1} + \frac{-2}{4} + \frac{-1}{5} + \frac{-1}{6}\right) = \pi.\frac{15}{47}\left(\frac{47}{15}\right) = \pi$$

Equation 8.44

Equation 8.44 has a number of homologs of which the most interesting may be:-

$$I\pi = A\int_0^{IC2} \frac{1}{a}.du + B\int_0^{IC2} \frac{1}{b}.dv + C\int_0^{IC2} \frac{1}{c}.dw + D\int_0^{IC2} \frac{1}{d}.dy$$
$$= \pi$$

Equation 8.45

The Integral of G in terms of the Numerator and Denominator Polynomials

At Equation 8.29 we have seen that:-

$$G_5 = \frac{F_5}{16^5} = \frac{120k^2 + 151k + 47}{512k^4 + 1024k^3 + 712k^2 + 194k + 15} = \frac{Numer}{Denom}$$
Equation 8.29

Further to this the Integral of G5, Ig5, is:-

$$Ig_5 = \frac{Numer}{Denom} = \int_0^1 \frac{Numer}{Denom} . dx$$
$$= \int_0^1 \frac{3802}{466785} . dx = \frac{3802}{466785}$$
$$= 0.008145077498206$$
Equation 8.46

A Note on the Square Root of Two, Ψ, the Pythagoras's Constant

I discovered that four times the Integral of BBP-for-π F, Iq, is approximately the Square Root of Two, Ψ:-

$$Ipc1 = 4 . Iq = 4 \int_0^\infty \frac{1}{16^k}$$
$$\cdot \left(\frac{A}{8k + a} - \frac{B}{8k + b} - \frac{C}{8k + c} - \frac{D}{8k + d} \right) . dk$$
Equation 8.47

or:-

$$Ipc2 = 4 \int_0^{m2} \frac{1}{16^k} \frac{120k^2 + 151k + 47}{512k^4 + 1024k^3 + 712k^2 + 194k + 15} . dk$$
Equation 8.48

The respective *percentage* specific defects are given by:-

$$PSD(\sqrt{2}, Ipc1) = -0.029257110663604$$
Equation 8.49

and:-

$$PSD\left(\sqrt{2}, Ipc2\right) = -0.029272106074964$$
Equation 8.50

In each case the error is about one part in 3333: Too much to be numerical error, and too small to indicate coincidence.

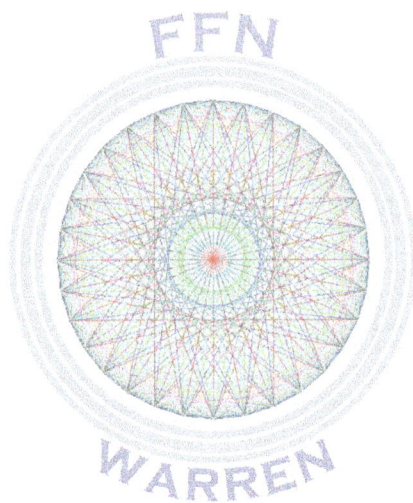

CHAPTER NINE
SOME BBP-TYPE FORMULAE
AS APPLIED TO THE DETERMINATION OF
FAMOUS CONSTANTS

Fiducial Values

Working to fifteen-digit accuracy on light machinery, we may conveniently define these fiducial values, arising from MathCad® Express®:-

The Ludolphine Constant, π

$$\pi_{fido} = \pi = 3.14159265358979$$
Equation 9.1

The Minor Ratio of Phidias, ϕ

$$\phi_{fido} = \frac{-1 + \sqrt{5}}{2} = 0.618033988749895$$
Equation 9.2

The Major Ratio of Phidias, Φ

$$\Phi_{fido} = \phi_{fido} + 1 = 1.618033988749895$$
Equation 9.3

The Square Root of Two, Pythagoras's Constant, Ψ

$$\Psi_{fido} = \sqrt{2} = 1.4142135623731$$
Equation 9.4

The Square Root of Three, Theodorus' Constant, χ

$$\chi_{fido} = \sqrt{3} = 1.73205080756888$$
Equation 9.5

The Square Root of Five

$$Sqr5_{fido} = \sqrt{(\Psi_{fido}\cdot\chi_{fido})^2 - 1} = 2.23606797749979$$
Equation 9.6

The $\pi\Psi$ Product

$$\pi_{fido}\Psi_{fido} = 4.44288293815837$$
Equation 9.7

The Napierian Base (Euler's Number), e

$$e_{fido} = e = 2.71828182845905$$
Equation 9.8

Formulae of BBP Type[9.1]

Formulae of BBP-type have been developed from the 1997 paper by David Bailey, Peter Borwein and Simon Plouffe which first proposed the structure in the context of computing Pi to high accuracy, and in particular extracting particular (hexadecimal) digits from within the mantissa.

In this century numerous "BBP-type" formulae have been presented to compute other well-known constants or their interior digits, both by the original team and by other workers.

The general BBP-type formula is given by:-

$$\alpha = \frac{1}{r}\sum_{k=0}^{m2}\frac{1}{\beta^k}\sum_{j=1}^{m3}\frac{a_j}{(Mk+j)^s}$$
Equation 9.9

where α is some Constant; β is the "Base" (relevant for digit extraction); and k, j, m2, m3, a_j, M and s are all suitable integers.

For the choices:-

$$m3 = 10 \qquad m2 = 10 \qquad k = 0...m3 \qquad j = 1...m2$$
$$r = 1 \qquad s = 1 \qquad M = 8 \qquad a_0 = 1$$
$$a_1 = 4 \quad a_2 = 0 \quad a_3 = 0 \quad a_4 = -2 \quad a_5 = -1$$
$$a_6 = -1 \quad a_7 = 0 \quad a_8 = 0 \quad a_9 = 0 \quad a_{10} = 0$$

the outcome of Equation 9.9 is 3.14159265358979 and hence:-

$$PSD(\pi_{fido}, \alpha) = 0$$
Equation 9.10

The Function Pf() for Determining a BBP Constant's Value

Following the similar proposals of several I define the function:-

$$Pf(m3, m2, \beta, r, s, M, a) = \frac{1}{r} \sum_{k=0}^{m2} \frac{1}{\beta^k} \sum_{j=1}^{m3} \frac{a_j}{(Mk + j)^s}$$
Equation 9.11

For example, the classical BBP formula for Pi is represented by:-

$$Pf(10,10,16,1,1,8,a) = 3.14159265358979$$
Equation 9.12

Since Pf(6,10,16,1,1,8,a) produced the same fifteen-figure result we could have saved time by using that.

a is a six or ten figure integer array.

The Case of Pi

As we saw earlier, the BBP formulation for π is:-

$$\pi = \sum_{i=0}^{\infty} \frac{1}{16^i} \left(\frac{4}{8i+1} - \frac{2}{8i+4} - \frac{1}{8i+5} - \frac{1}{8i+6} \right)$$

$$= \frac{1}{r_{con}} \sum_{k=0}^{\infty} \left(\frac{1}{\beta^k} \times \sum_{j=0}^{4} \frac{n_j}{M.j + d_j} \right)$$

$$= \sum_{k=1}^{\infty} \frac{120k^2 + 151k + 47}{\beta^{ck}.512k^4 + 1024k^3 + 712k^2 + 194k + 15}$$

Equation 9.13

where "infinity" is about ten for tolerable precision of thirteen to fifteen digits.

The Case of ln(2)

For this calculation m1 = 48.
Now:-

$$\ln(2)_{est} = \int_{k=1}^{m1} \frac{1}{k \cdot 2^k} = 0.693147180559945$$

Equation 9.14

for which:-

$$PSD(\ln(2), \ln(2)_{est}) = 0.000000000000032$$

Equation 9.15

Given that $a_0 = 0$ and $a_1 = 1$ then an Estimate of ln(2), call it ln(2)$_{est}$, is:-

$$\ln(2)_{est} = \frac{2}{3} \cdot Pf(1,1,9,1,1,1,a) = 0.703703703703704$$

Equation 9.16

This is essentially an uniterated estimate, for which:-

$$PSD(\ln(2), \ln(2)_{est}) = -1.52298435885299$$
Equation 9.17

A much better Pf() estimate is achieved using the following formulation:-

$$\ln(2)_{est} = \frac{1}{2}Pf(m3, m2, \beta, r, s, M, a)$$
$$= \frac{1}{2}Pf(1, 48, 2, 1, 1, 1, a)$$
$$= 0.693147180559945$$
Equation 9.18

for which the PSD is:-

$$PSD(\ln(2), \ln(2)_{est}) = 0.000000000000032$$
Equation 9.19

Relationship to Euler's Constant

$$\Psi = \sqrt{2} = e^{\frac{\ln(2)}{2}}$$
Equation 9.20

The Case of ln(3)

Once again, m1 = 48.

$$\ln(3) = 2.\ln(2) + \ln\left(\frac{3}{4}\right) = 1.09861228866811$$
Equation 9.21

from which it follows that:-

$$\ln(3) = 2\sum_{k=1}^{m1}\frac{1}{k \cdot 2^k} - \sum_{k=1}^{m1}\frac{1}{k \cdot 4^k}$$

$$= \frac{1}{2} \sum_{k=0}^{m1} \frac{1}{4^k} \left(\frac{2}{2k+1} + \frac{1}{2k+2} \right) - \frac{1}{4} \frac{1}{4^k} \left(\frac{2}{2k+1} \right.$$

$$\left. + \frac{1}{2k+2} \right) \sum_{k=0}^{m1} \frac{1}{4^k} \left(\frac{1}{2k+2} \right)$$

$$= \sum_{k=0}^{m1} \frac{1}{4^k} \left(\frac{1}{2k+1} \right)$$

$$= 1.09861228866811$$

Equation 9.22

Correspondingly, with again $a_0 = 0$ and $a_1 = 1$, then in terms of the function Pf():-

$$\ln(3)_{est} = Pf(m3, m2, \beta, r, s, M, a)$$
$$= Pf(1,48,4,1,1,2,a)$$
$$= 1.09861228866811$$

Equation 9.23

for which:-

$$PSD(\ln(3), \ln(3)_{est}) = 0.000000000000081$$

Equation 9.24

Relationship to Euler's Constant

$$\chi = \sqrt{3} = e^{\frac{\ln(3)}{2}}$$

Equation 9.25

The Case of $\pi \Psi^{9.2}$

$$\pi \Psi_{est1} = \sum_{k=0}^{m1} \frac{1}{(-8)^k} \left(\frac{4}{6k+1} + \frac{1}{6k+3} + \frac{1}{6k+5} \right)$$

Equation 9.26

or:-

$$\pi\Psi_{est2} = \sum_{k=0}^{m1} \frac{1}{(-8)^k}\left(\frac{4}{6k+1} + \frac{0}{6k+2} + \frac{1}{6k+3} + \frac{0}{6k+4}\right.$$
$$\left. + \frac{1}{6k+5}\right)$$

Equation 9.27

These First and Second BBP-type Estimates of $\pi\Psi$ both give the numerical value 4.44288293815837 so both are "exact" to fifteen figures for $\pi\Psi$.

Correspondingly, for Equation 9.27:-

$$\pi\Psi_{est2} = Pf(m3, m2, \beta, r, s, M, a)$$
$$= Pf(5, 48, -8, 1, 1, 6, a)$$
$$= 4.44288293815837$$

Equation 9.28

which is, therefore, also "exact".

Alternate Form

The following BBP-type Estimator for $\pi\Psi$ is sometimes seen in literature:-

$$\pi\Psi_{est3} = \frac{1}{8}\sum_{k=0}^{m1} \frac{1}{(64)^k}\left(\frac{32}{12k+1} + \frac{8}{12k+3} + \frac{8}{12k+5} - \frac{4}{12k+7}\right.$$
$$\left. - \frac{1}{12k+9} - \frac{1}{12k+11}\right)$$

Equation 9.29

As it stands, Equation 9.29 has a PSD:-

$$PSD(\pi\Psi, \pi\Psi3) = 0.00000000000002$$

Equation 9.30

Please note that Formula 8 in Weisstein is wrong, and consequentially so is the copy in Wikipedia "Square root of 2".

For this Third Estimate of $\pi\Psi$ the Pf() function forms are:-

$$\pi\Psi3 = Pf(m3, m2, \beta, r, s, M, a)$$
$$= Pf(11,48,64,8,1,12, a)$$
$$= 4.44288293815837$$
Equation 9.31

for which:-

$$PSD(\pi\Psi, \pi\Psi3) = 0.00000000000002$$
Equation 9.32

PseudoBBP Formula for e the Euler Constant[9.3]

As published by JacobTDC:-

$$e_{pseudoBBP} = \sum_{k=0}^{48} \left(\frac{1}{4}\right)^k \left(\frac{4k+3}{(4k+2)\cdot(2k)!}\right) = 1.64872127070013$$

In his otherwise excellent post, JacobTDC forgot to square the RHS.

Accordingly, I offer the adjusted formula below:-

$$e_{pseudoBBP} = \left[\sum_{k=0}^{48} \left(\frac{1}{4}\right)^k \left(\frac{4k+3}{(4k+2)\cdot(2k)!}\right)\right]^2$$
$$= 2.71828182845905$$
Equation 9.33

The PSD for Equation 9.33 is:-

$$PSD(e_{fido}, e_{pseudoBBP}) = -0.000000000000016$$
Equation 9.34

It is not yet known whether there exists a true BBP formula for e (30 September 2023).

<u>The Case of the Minor Ratio of Phidias, ϕ</u>

Firstly we will say a few words about the Major Ratio of Phidias, Φ which is unity plus the Minor Ratio, ϕ:-

$$\Phi = \frac{1 + \sqrt{5}}{2} = e^{-\ln(\phi)} = 1.61803398874989$$

Equation 9.35

so that:-

$$\phi = \frac{\sqrt{5} - 1}{2} = 0.618033988749895$$

Equation 9.36

Allow that:-

m2 = 48	m3 = 2	k = 0...m2	j = 1...m3
r = -√5	β = 5	s = 1	M = 2
a₀ = 0	a₁ = 1	a₂ = 0	

then:-

$$\ln(\phi) = \frac{1}{r} \sum_{k=0}^{m2} \frac{1}{\beta^k} \sum_{j=1}^{m3} \frac{a_j}{(Mk + j)^s}$$
$$= Pf(m3, m2, \beta, r, s, M, a)$$
$$= Pf(2, 48, 5, -\sqrt{5}, 1, 2, a)$$
$$= -0.481211825059603$$

Equation 9.37

Hence:-

$$\phi = e^{Pf(2,48,5,-\sqrt{5},1,2,a)} = 0.618033988749895$$
$$\textbf{Equation 9.38}$$

from which:-

$$\ln(\phi) = -e^{\ln[-Pf(2,48,5,-\sqrt{5},1,2,a)]} = -0.481211825059603$$
$$\textbf{Equation 9.39}$$

or:-

$$-\ln(\phi) = e^{\ln[-Pf(2,48,5,-\sqrt{5},1,2,a)]} = 0.481211825059603$$
$$= -Pf(2,48,5,-\sqrt{5},1,2,a)$$
$$\textbf{Equation 9.40}$$

Synthesis

Pf() for Pi

$a_0 = 1$
$a_1 = 4$ $a_2 = 0$ $a_3 = 0$ $a_4 = -2$ $a_5 = -1$
$a_6 = -1$ $a_7 = 0$ $a_8 = 0$ $a_9 = 0$ $a_{10} = 0$

$$\pi = Pf(6,10,16,1,1,8,a) = 3.14159265358979$$
$$\textbf{Equation 9.41}$$

$$PSD\big(\pi, Pf(6,10,16,1,1,8,a)\big) = 0$$
$$\textbf{Equation 9.42}$$

Pf() for PiPhi (Product $\pi\Psi$)

$a_0 = 0$
$a_1 = 4$ $a_2 = 0$ $a_3 = 1$ $a_4 = 0$ $a_5 = 1$
$a_6 = 0$ $a_7 = 0$ $a_8 = 0$ $a_9 = 0$ $a_{10} = 0$

$$\pi\Psi = Pf(5,48,-8,1,1,6,a) = 4.44288293815837$$
$$\textbf{Equation 9.43}$$

$$PSD\big(\pi, Pf(5,48,-8,1,1,6,a)\big) = 0$$
$$\textbf{Equation 9.44}$$

The Pythagoras's Constant, Ψ, and the Euler's Number, e

Note that by transposition:-

$$2 = \frac{\sqrt{5} - 1}{\phi}$$

Equation 9.45

and that:-

$$\Psi = \sqrt{2} = e^{\frac{\phi.\ln(2)}{\sqrt{5}-1}} = 1.41421356237309$$

Equation 9.46

The relevant PSD is:-

$$PSD\left(\Psi, e^{\frac{\phi.\ln(2)}{\sqrt{5}-1}}\right) = 0.000000000000016$$

Equation 9.47

The Composite Target Constant, T

Firstly we may check whether there is any discrepancy between the product of the separate constants π and Ψ ($\pi \times \Psi$) on the one hand, and the made-earlier composite constant $\pi\Psi$. Unfortunatly, we cannot assume the two outcomes identical just because they are mathematical identities. Any discrepancy reflects the numerical error between the two MathCad® Express® formations:-

$$e^{\frac{\pi_{fido} \times \Psi_{fido}}{\phi_{fido}}} = 1324.4275429116$$

Equation 9.48

and:-

$$e^{\frac{\pi\Psi_{BBP}}{\phi_{BBP}}} = 1324.4275429116$$

Equation 9.49

The subscript "fido" denotes the fifteen-figure Fiducial Value as computed by MathCad® Express® whilst the subscript "BBP" represents values computed via appropriate Bailey-Borwein-Plouffe formulae, but also negotiated by MathCad® Express® mechanics.

The relative PSD is:

$$PSD\left(e^{\frac{\pi_{fido}\times\Psi_{fido}}{\phi_{fido}}}, e^{\frac{\pi\Psi_{BBP}}{\phi_{BBP}}}\right) = 0.000000000000172$$

Equation 9.50

So computational error is 1.72×10^{-13}
Moving forward:-

$$exp\left(\frac{Pf(5,n,-8,1,1,6,a)}{e^{Pf(2,n,5,-\sqrt{5},1,2,a)}}\right)$$
$$= \exp\left(Pf(5,n,-8,1,1,6,a).e^{-Pf\left(2,n,5,-\sqrt{5},1,2,a\right)}\right)$$
$$= 1324.4275429116$$

Equation 9.51

But n = 19 is nescessary and sufficient, whereupon:-

$$PSD\left(e^{\frac{\pi_{fido}\times\Psi_{fido}}{\phi_{fido}}}, e^{e^{\frac{Pf(5,19,-8,1,1,6,a)}{Pf(2,19,5,-\sqrt{5},1,2,a)}}}\right) = 0$$

Equation 9.52

Simplification and the Taylor Series

Given that further:-

$$a\phi_0 = 0 \qquad a\phi_1 = 1 \qquad a\phi_2 = 0$$

and that n = 19 we may locally define:-

$$f = Pf(5, n, -8,1,1,6, a)$$
Equation 9.53

$$g = Pf(2, n, 5, -\sqrt{5}, 1,2, a\phi)$$
Equation 9.54

so that:-

$$exp\left(\frac{f}{e^g}\right) = exp\left(\frac{\pi^\Psi}{\phi}\right)$$

Equation 9.55

and:-

$$PSD\left(e^{\frac{\pi_{fido}\times\Psi_{fido}}{\phi_{fido}}}, e^{\frac{f}{e^g}}\right) = 0$$

Equation 9.56

With regard to the Taylor Series Expansion, adequate n is 38 so that:-

$$TaySer = \sum_{k=0}^{n}\frac{1}{k!}f^k e^{-kg} = 1324.4275429116$$

Equation 9.57

$$PSD\left(e^{\frac{\pi_{fido}\times\Psi_{fido}}{\phi_{fido}}}, TaySer\right) = 0$$

Equation 9.58

CHAPTER TEN
EPILOG

Then Jonah prayed unto the LORD his God out of
the fish's belly,
And said, I cried by reason of mine affliction unto
the LORD, and he heard me; out of the belly of
hell cried I, and thou heardest my voice.
For thou hadst cast me into the deep, in the midst
of the seas; and the floods compassed me about:
all thy billows and thy waves passed over me.
Then I said, I am cast out of thy sight; yet I will
look again toward thy holy temple.
The waters compassed me about, even to the soul:
the depth closed me round about, the weeds were
wrapped about my head.
I went down to the bottoms of the mountains; the
earth with her bars was about me for ever: yet hast
thou brought up my life from corruption, O LORD
my God.

- KJV Jonah Ch2: 1-6

God made slugs free so they husband their portion as they please.

God made men free so they question Creation, and despoil the sea, in so far as they are able.

Man is an ape. But an ape more like a chimp or a bonobo than the stately orang, and man has none of the gravitas of the gorilla.

Fixated upon strife and sex, Man prances about a bit, cackles, grimaces and is gone.

Man plays under God. Man struggles under God to descry the Reality of Nature or the Nature of Reality and the relation of Reason, if any, to either.

Man is finite in time and understanding and perforce reduces everything to caricature and stereotype.

Man judges, though Judgment is neither his aptitude nor his province.

Like man the slug or the worm gets what he is given, but also like the slug man endeavors to garner what he can, aware that his time is short and his gruel thin.

<u>Chondrites</u>

The diagram Figure 10.2 is a caricature, and there is more evidence that the burrowing organism "intends" a 35° Y-branch rather than systematic construction of regular dendroids.[10.1,10.2,10.3]

Chandra and Putterer offer independent estimates, approximations, of the interangles of branching Chondrites burrows. Chandra cites 25, 30 and 40 degrees and Putterer 25, 30, 40, 30, and 40 degrees.

Such branching consistency is very suggestive of systematic behavior by the Chondrites organism, but much more goniometric work would be required before tenable interpretations became feasible.

Very provisionally, though, I shall propose that the bifurcative interangle is ϕ radians and compute whether the available data, rough and sparse as it is, supports such a speculation. Accordingly, let a function Degtophi() represent the conversion of an Angle in Degrees to an angle in (decimal) Fractions of ϕ:-

$$DegTophi(deg) = \frac{deg}{\frac{\phi}{\pi} \cdot 180} = \frac{\pi}{180} \cdot \frac{deg}{\phi}$$

Equation 10.1

and the Arithmetic Mean of n such Angles be:-

$$\mu_\alpha = \frac{1}{n} \sum_{i=1}^{n} DegTophi(deg)$$

Equation 10.2

Then for the Chandra offering with α_i = 25, 30 and 40 degrees μ_α = 0.894267316240647 and for Putterer's α_i = 25, 30, 40, 30 and 40 degrees μ_α = 0.931920676924464

In my opinion Chondrites was likely an annelid (worm) because some of those creatures are known to divide in adulthood to bear two or more clones who then part in their separate ways. This is a process of asexual fission.

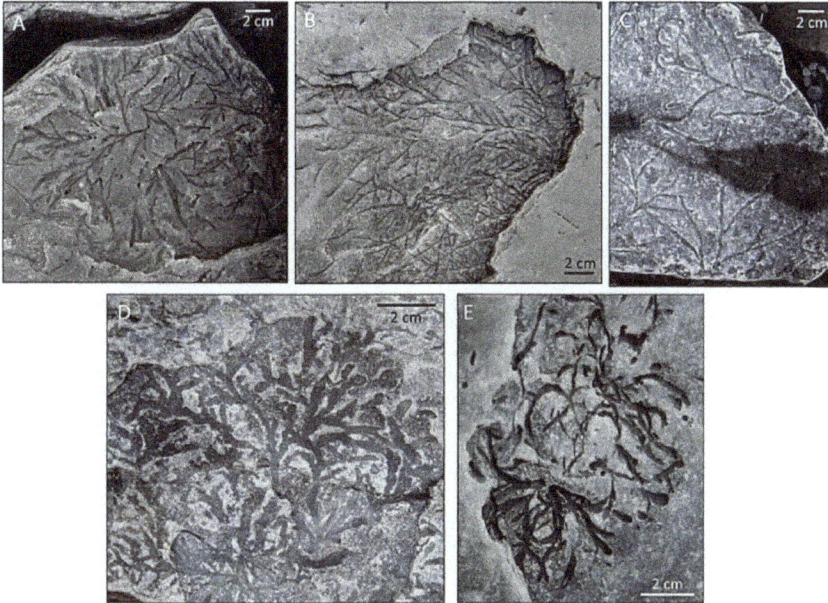

Fig. 2.3. *Chondrites* preserved in silty fine-grained sandstones. A-C) *Chondrites targionii*; D-E) *Chondrites recurvus*

Photo: Robyn Rebecca Reynolds

Figure 10.1
Chondrites in a Newfoundland Arenite

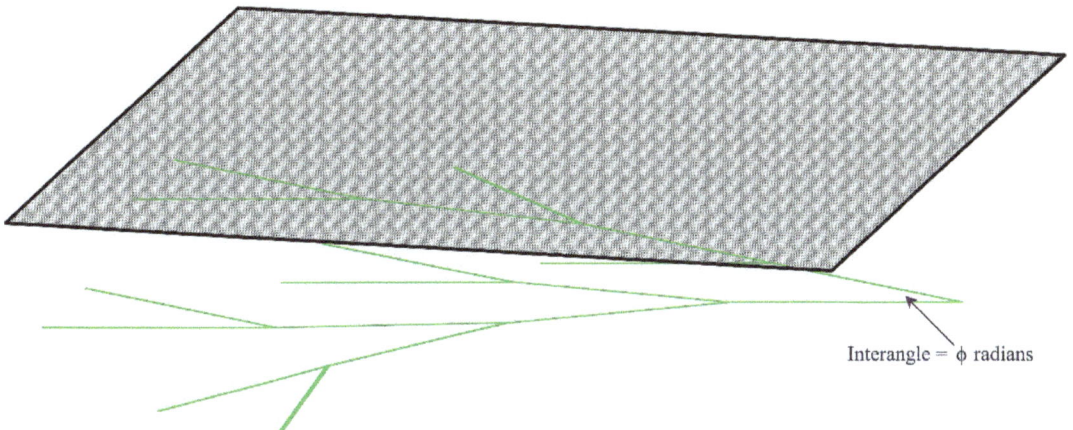

Interangle = φ radians

Figure 10.2
Schematic Diagram of Chondrites Propagations
Slightly Below a Sediment-Water Interface

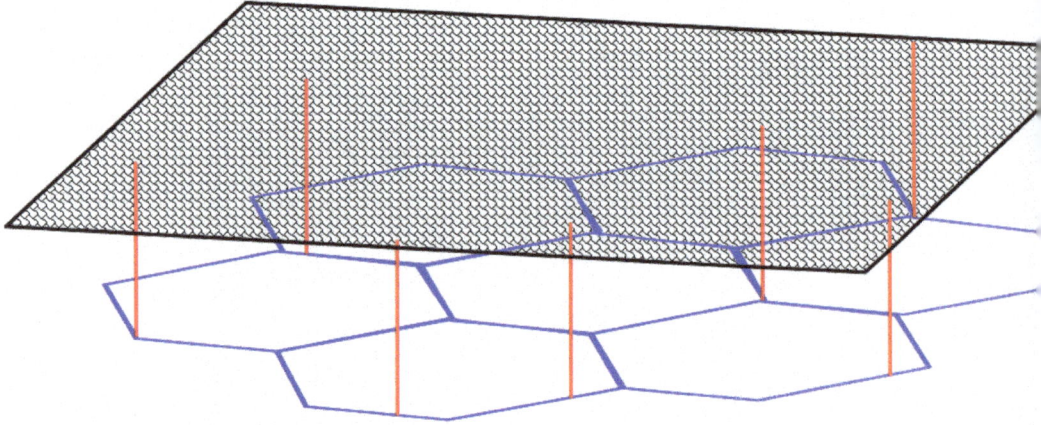

Figure 10.3
Schematic Diagram of Paleodictyon-type Reticulation
Slightly Below or at a Sediment-Water Interface

          Paleodictyon is a sedimentary trace fossil found in arenaceous facies of Early Cambrian (538.8 million years BP) to Recent dates. Context infers that the genitive agent (I hesitate to call it an organism) inhabited abyssal to bathyal marine environments. Modern *Paleodictyon nodosum* burrows have been retrieved using the submersible research craft "Alvin" from a depth of 3,500 meters. The digger of the burrows (if any) was not detected.

          On the contrary, Ophiomorpha burrows tend to be littoral though abyssal examples have been found. Ophiomorpha can be callianassid (ghost shrimp) burrows. Unlike Paleodictyon, Ophiomorpha burrows are actively backfilled. Ophiomorpha emerged in the Early Permian (298.9 million years BP) and is still with us. Ophiomorpha is reticular but much rougher both in plan and section than Paleodictyon.

          It does not take much imagination to wonder at the helmsmanship of whatever or whoever dug these burrows in the ooze in the pitch-dark abyss at a chilly 2°C and a pressure of 76MPa (11000 atmospheres).

          The blue hexagonal cells are diagrammatically-stylised Paleodictyon or Ophiomorpha burrows, controversially used for cultivating food bacteria, and the red risers (not always at hexagon vertices) "prove" that the blue burrows were dug by organisms rather than being contraction cracks or other thermodynamic stress phenomena such as Rayleigh-Bénard convection cells.[10.4]

The gray shingled surface represents the silt or sand-water interface, that is the sea-bed. The red risers terminate at this plane.

Paleodictyon and Ophiomorpha burrows are precise hexagonal meshes upon an ancient and indurated lithological bedding-plane, very reminiscent of the open-weave galvanised steel chicken wire you may readily find in rural hardware stores. It is obvious that statistical goniometry could be used to assess the quality and consistency of the pattern. As in geology generally it is the interangles that matter more than the shape.

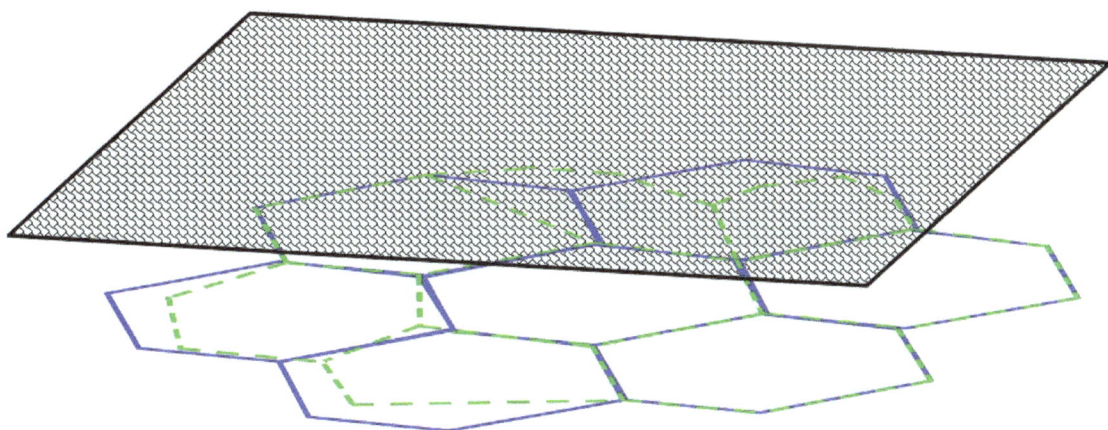

Figure 10.4
Perturbed Paleodictyon-type Reticulation
Slightly Below or at a Sediment-Water Interface

Some have postulated that Paleodictyon or Ophiomorpha cells are hot-rock convectional phenomena due to the buoyant ascent of dense plumes in a thin but mobile layer trapped at high pressures and temperatures below overburden. This is known as Rayleigh–Bénard convection. Such cellular convections are controlled by the Froude Number[10.5], a dimensionless parameter which is essentially the ratio of kinetic to potential energy in a fluid system and in context may be quoted as:-

$$Fr = \frac{v}{\sqrt{g'.h}}$$

Equation 10.1

where Fr is the Froude Number; v is the Velocity of the Moving Fluid; h is the "Characteristic Length" of the Solidifying Magma Column; g is the (Standard) Acceleration Due to Gravity; g′ is the Boussinesq Adjusted Acceleration due to Gravity; and $\Delta\rho$ is the Differential Density defined by:-

$$g' = g\frac{\rho_1 - \rho_2}{\rho_1} = g.\Delta\rho$$

Equation 10.2

where ρ is the Gravimetric Density, ρ_1 and ρ_2 being the respective Densities of two Interfacing Fluid Bodies. In some literature you will find that the Froude Number is contextually defined as the ratio of buoyancy to inertia.

As a geologist and hydraulician I am disturbed by the wildly hypercritical Froude Numbers (~ 1300) sometimes proposed. This issue is explored in Appendix C.

Paleodictyon and Ophiomorpha cells are always found in unmelted sedimentary facies.

Some say that these precisely hexagonal cells are tensional cracks like swarming geological faults, but if two convergent lines are visible upon a Euclidian plane, then those lines cannot be coplanar: Tensional cracks are coplanar.

Paleodictyon is crossed by three tensional cracks manifestly at different angles to any of the hexagon sides in the Figure 10.5 photograph.

Any two Euclidian planes that trace convergent lines upon a third surface describe three plane pairs none of which can be coplanar.

At least two coplanar sets of tensional cracks are shown on the photograph.

No such lines parallel a hexagon side shown.

Therefore, the hexagon sides are not tensional fissures.

Convection phenomena are controlled by the Froude Number and for such convection cells to form in for example basaltic lava the Froude numbers calculated are very high.

But Paleodictyon or other hexagonally-reticular trace fossils are geometrically far too regular credibly to be convection cells, which are typically in plan an assortment of triangular, pentagonal and especially sub-regular hexagonal networks.

The Wikipedia photograph of Figure 10.6 shows laboratory formation of Rayleigh-Bénard cells in roiling, viscous fluid, not impossibly Everton toffee.[10.7]

For a geological comparison my photograph Figure 10.7 of part of the Am Buachaille formation (NM 32561 35136: 56° 14′ 27″ N 6° 20′ 24′′ W) shows the heterogeneity of the basaltic prism sections.

It is thought that the curvature of the prismatic columns is caused by a differential thermal gradient at the time of gradual solidification, or in rare cases by slumping into volcanic ash.[10.8]

Another example of prismatic heteromorphism is shown by basaltic columns at the Giant's Causeway (NR 13319 03355: 55° 14′ 27″ N 6° 30′ 43″ W) on the coast of County Antrim, part of which is shown in my Figure 10.8

On the contrary, Paleodictyon or Ophiomorpha cells are rigidly regular like the open weave of chicken wire or indeed the honeycomb worked in wax by melittids.

My favorite theory of Paleodictyon formation is the sub-surface burrowing of fodinichnial agents likely to be annelids, but there are real problems with this theory also.

My diagram Figure 10.9 illustrates the suggested behavior starting with Step 1 and after precisely a unit step turning the helm precisely $-\pi/3$ radians to larboard or $+\pi/3$ to starboard over the course of only four hexagons.

Repeated steps are shown by the four double edges.

Photo: Falconaumanni

Figure 10.5
Paleodictyon imperfectum? in Flysch Sandstone at Algeciras,
Spain[10.6]

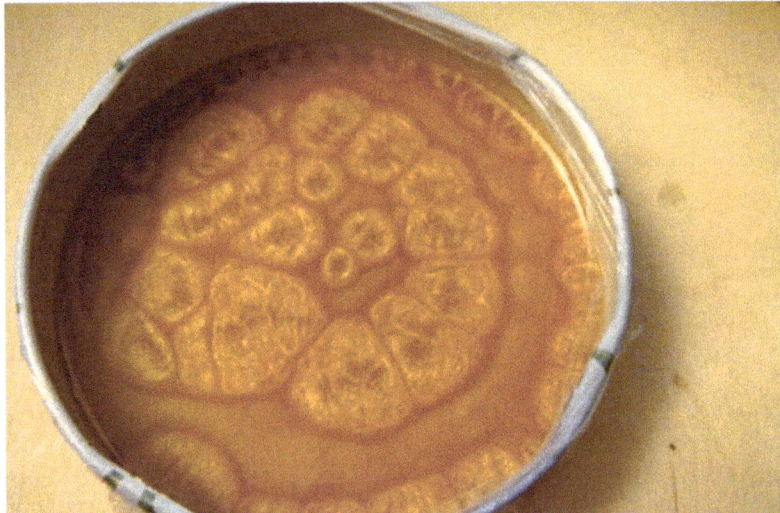

RayleighBenardConvectionCellsScreenshot 2023-10-15 162643

Figure 10.6
Roiling Viscous Fluid showing Rayleigh-Bénard Cellular
Convection

Photo: James R Warren

Figure 10.7
Sub-Hexagonal Basaltic Columniation in the
Tertiary Igneous Province, Staffa, Scotland

Photo: James R Warren

Figure 10.8
Sub-Hexagonal Basaltic Columniation in the
Tertiary Igneous Province, The Giant's Causeway, Ireland

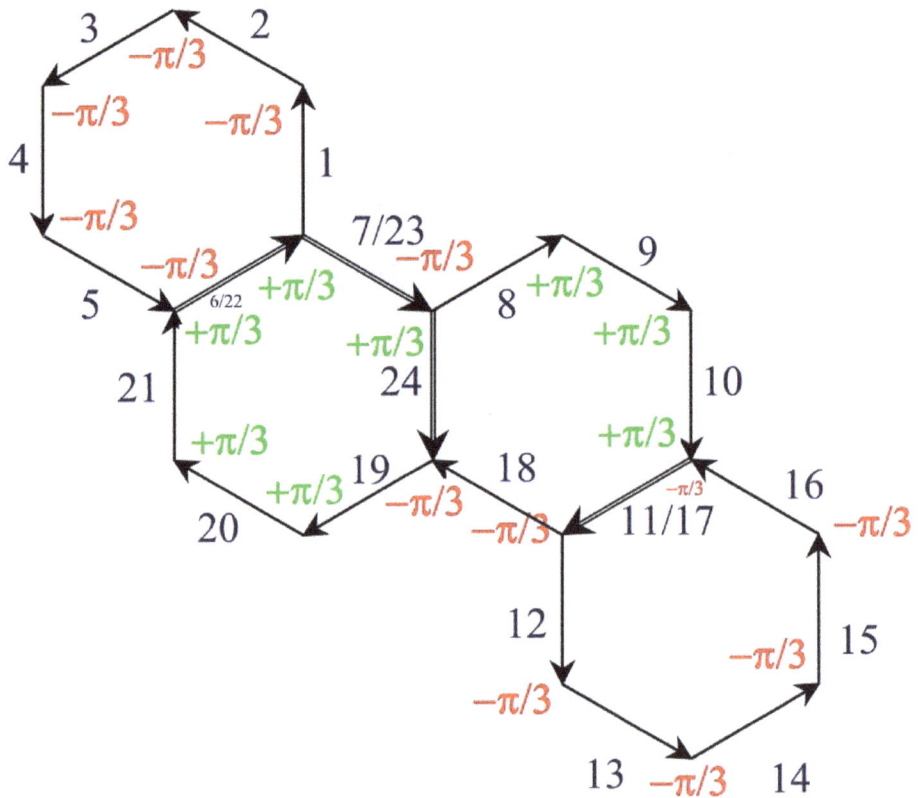

Figure 10.9
Double-Traverse in the Continuous Excavation of a Hexagonal Lattice

Each of the 24 steps are precisely equal, but 4 of the 24 are repeated, i.e. 1 in 6. This is an inefficient strategy, and inefficiency is unusual both in plants and in invertebrates of any phylum, though it is common enough in chordates and especially mammals. The attraction from a mathematical point of view is the exact and apparently programmatic cybernation of whatever agent it may be that formed the Paleodictyon burrow.

The choice of turn angles is (at least to the extent of our diagram) restricted to $+\pi/3$ radians ($+60°$) in green, or $-\pi/3$ radians ($-60°$) in red, as previously alluded to.

These efficiency issues may be clarified by more work, possibly implicating the Euler Circuit or other aspects of graph theory.

In 1773AD, Joseph Louis Lagrange showed that:-

$$\eta = \frac{\pi}{2\sqrt{3}} \approx 0.906899682117109$$

Equation 10.3

where η is the Packing Density, otherwise known as the Packing Fraction.

The hexagon is one of only three polygons that permit close-packing. In particular, the limit polygon, the circle, leaves a 9.3% areal waste. Meanwhile the square, whilst permitting wasteless packing, exposes any structure of packed squares to lateral shear crushing.

To be explicit, the honeycomb structure of packed hexagons permits the maximum of physical structural strength to be combined with the maximum of contained area.

This led McMenamin[10.9] to propose that Paleodictyon was a benthic or very shallow sub-benthic swarm of nests (with riser burrows) where the organism laid its eggs and reared its young. This is substantiated by the lack of corporeal fossils within the hexagons, since soft-bodied forms would unlikely survive the induration.

We were not present when Chondrites and Paleodictyon worked and we must take more or less educated guesses as to their methods, or indeed as to their ancient existence.

We may exercise mathematics and the "known" laws of physics to guide our inferences, including our knowledge of the Four Famous Numbers, whose properties we decide for ourselves, but which are strangely mirrored in the Natural World, even the remotest parts of Nature in space and time.

BIBLIOGRAPHY AND REFERENCES

Chapter One

1.0   Epigraph
"Sartor Resartus"
Thomas Carlyle
Oxford World's Classics
Oxford University Press of New York   1987
ISBN 978-0-19-954037-2
pp 273
p14 line -13 to line -7

1.1   Wikipedia contributors. (2022, May 30).
Nereites.
In Wikipedia, The Free Encyclopedia.
Retrieved 10:13, September 11, 2023, from
https://en.wikipedia.org/w/index.php?title=Nereites&oldid=1090681444

1.2   English: Nereites irregularis
(formerly Helminthoida labyrinthica) trace fossil
Richard Ford

https://commons.wikimedia.org/wiki/File:Nereites_irregularis.jpg

https://upload.wikimedia.org/wikipedia/commons/7/79/
Nereites_irregularis.jpg

Richdebtomdom, CC BY-SA 3.0
<https://creativecommons.org/licenses/by-sa/3.0>, via Wikimedia Commons

<a title="Richdebtomdom, CC BY-SA 3.0
<https://creativecommons.org/licenses/by-sa/3.0>,
via Wikimedia Commons"
href="https://commons.wikimedia.org/wiki/File:Nereites_irregularis.jpg">
<img width="512" alt="Nereites irregularis" src=

"https://upload.wikimedia.org/wikipedia/commons/thumb/7/7
9/
        Nereites_irregularis.jpg/512px-
Nereites_irregularis.jpg"></a>

    1.3   "Researches: Volume Three"
            James R Warren
            Midland Tutorial Productions of Bloxwich
            1 October 2022
            ISBN 978 1 915750 02 0
            pp 288

            SEARCHPT.DOC   25 August 2009
            "The Characteristics of Some Search Patterns"
            p 11-40

    1.4   "Pi and Phi: A Coy Romance of Two Numbers"
            Second Edition
            James R Warren
            Midland Tutorial Productions of Bloxwich
            1 September 2023
            ISBN 978 1 915750 06 8
            pp 424

Chapter Two

      (No References)

Chapter Three

      (No References)

Chapter Four

4.1   Wikipedia contributors. (2023, September 17).
       Approximations of π.
       In *Wikipedia, The Free Encyclopedia*.
       Retrieved 10:25, September 21, 2023, from
https://en.wikipedia.org/w/index.php?title=Approximations_of_%CF
%80&oldid=1175733173

Chapter Five

5.1   "Handbook of Mathematical Functions
       with
       Formulas, Graphs and Mathematical Tables"
       Edited by Milton Abramowitz and Irene A Stegun
       United States Department of Commerce
       Washington DC
       National Bureau of Standards
       Applied Mathematics Series . 55
       Issued June 1964
       Tenth Printing December 1972 with Corrections
       1046pp
       Simpsons Rule   p886 25.4.6 Extended Simpsons Rule
       Rule 18   p886 25.4.18

5.2   Wikipedia contributors. (2023, July 19).
       Mathematical constant.
       In Wikipedia, The Free Encyclopedia.
       Retrieved 10:15, August 18, 2023, from
https://en.wikipedia.org/w/index.php?title=Mathematical_constant&o
ldid=1166115764

5.3   Wikipedia contributors. (2023, August 5).
Square root of 2.
In Wikipedia, The Free Encyclopedia.
Retrieved 10:17, August 18, 2023, from
https://en.wikipedia.org/w/index.php?title=Square_root_of_2&oldid=
1168832293

5.4   (OEIS A070197; Wells 1986, p.35; Guy 1990;
Conway and Guy 1996, pp. 181-182; Flannery 2006,
pp. 32-33)

5.5   The Online Encyclopedia of Integer Sequences
NJA Sloane
https://oeis.org/A070197

5.6   "The Square Root of 2,
A Dialogue Concerning A Number And A Sequence"
David Flannery
Copernicus Books, NY, 2006, pp. 32-33.

5.7   E.W. Weisstein, Aug. 30, 2008).
Weisstein, Eric W. "Pythagoras Constant."
From Mathworld--A Wolfram Web Resource.
https://mathworld.wolfram.com/PythagorassConstant.html

Chapter Six

6.1   University of the Pacific
Scholarly Commons
https://scholarlycommons.pacific.edu/euler-
works/853/

University of the PacificUniversity of the Pacific
Scholarly Commons
Euler Archive - All Works by Eneström Number
 Euler Archive
1862

"Meditatio in experimenta explosione tormentorum
nuper instituta"
Leonhard Euler

Follow this and additional works at:
https://scholarlycommons.pacific.edu/euler-works
Record Created: 2018-09-25
Recommended Citation
Euler, Leonhard,
"Meditatio in experimenta explosione tormentorum
nuper instituta"   (1862).
Euler Archive -
All Works by Eneström Number. 853.
https://scholarlycommons.pacific.edu/euler-w
(Facsimile of 19th century letterpress in Latin)

6.2   Wikipedia contributors. (2023, September 23).
E (mathematical constant).
In Wikipedia, The Free Encyclopedia.
Retrieved 10:27, September 25, 2023, from
https://en.wikipedia.org/w/index.php?title=E_(mathematical_constant
)&oldid=1176749350

6.3   "Improving the Convergence of Newton's Series
Approximation for e"
Harlan J Brothers
The Country School in Madison, Connecticut
The College Mathematics Journal
Vol.35, No.1, January 2004, pp34-39
Copyright: The Mathematical Association of America

6.4   Wallis, J. "Arithmetica Infinitorum".
Oxford, England, 1656.
Sondow, J. "A Faster Product for pi and a New
Integral for ln(pi/2)."
Amer. Math. Monthly 112, 729-734, 2005.
https://mathworld.wolfram.com/WallisFormula.html

6.5   Weisstein, Eric W.
        "Pippenger Product."
        From MathWorld--A Wolfram Web Resource.

https://mathworld.wolfram.com/PippengerProduct.html

6.6   "Exponential Function as an Infinite Product"
        pathfinder et al at 19:33 Nov 20, 2016
        https://math.stackexchange.com/questions/502160/exp
        onential-function-as-an-infinite-product

6.7   "Handbook of Mathematical Functions
        with
        Formulas, Graphs and Mathematical Tables"
        Edited by Milton Abramowitz and Irene A Stegun
        United States Department of Commerce
        Washington DC
        National Bureau of Standards
        Applied Mathematics Series . 55
        Issued June 1964
        Tenth Printing December 1972 with Corrections
        1046pp
        Simpsons Rule p886 25.4.6 Extended Simpsons Rule
        Rule 18   p886 25.4.18

6.8   Wikipedia contributors. (2023, July 19).
        Mathematical constant.
        In Wikipedia, The Free Encyclopedia.
        Retrieved 10:15, August 18, 2023, from
https://en.wikipedia.org/w/index.php?title=Mathematical_constant&o
ldid=1166115764

6.9   Wikipedia contributors. (2023, August 8).
        E (mathematical constant).
        In Wikipedia, The Free Encyclopedia.
        Retrieved 15:05, August 24, 2023, from
https://en.wikipedia.org/w/index.php?title=E_(mathematical_constant
)&oldid=1169366229

Chapter Seven

7.1   Wikipedia contributors. (2023, May 7).
Bailey–Borwein–Plouffe formula.
In Wikipedia, The Free Encyclopedia.
Retrieved 13:39, October 6, 2023, from
https://en.wikipedia.org/w/index.php?title=Bailey%E2
%80%93Borwein%E2%80%93Plouffe_formula&oldi
d=1153688764

7.2   Bailey, David; Borwein, Peter; Plouffe, Simon
"On the rapid computation of various polylogarithmic
constants."
Math. Comp.66(1997), no.218, 903–913.

Chapter Eight

8.1   "Handbook of Mathematical Formulas"
Hans-Jochen Bartsch
Translated by Herbert Liebscher
Edition for Academic Press of New York and London
A Subsidiary of Harcourt Brace Jovanovich
Library of Congress Catalog Number: 73-2088
ISBN 0-12-080050-0   (Hardback)
Copyright: VEB Publishing Leipzig 1974
pp525
Integration by Partial Fractions   pages 337-341

8.2   Wikipedia contributors. (2023, July 15).
Partial fraction decomposition.
In Wikipedia, The Free Encyclopedia.
Retrieved 14:12, October 13, 2023, from
https://en.wikipedia.org/w/index.php?title=Partial_fraction_de
composition&oldid=1165546716

Chapter Nine

9.1   "On the Rapid Computation of Various
Polylogarithmic Constants"
David H Bailey, Peter B Borwein and Simon Plouffe
Mathematics of Computation
V66, No 218, 1997, pp903-913

in
"A Compendium of BBP-Type Formulas
 for Mathematical Constants"
David H Bailey
April 8, 2023
https://www.davidbailey.com/dhbpapers/bbp-
formulas.pdf

9.2   Weisstein, Eric W. "Pythagoras's Constant."
        From MathWorld--A Wolfram Web Resource.
        https://mathworld.wolfram.com/PythagorassConstant.
        html
        in
        Wikipedia contributors. (2023, September 18).
        Square root of 2.
        In Wikipedia, The Free Encyclopedia.
        Retrieved 14:04, September 30, 2023, from
        https://en.wikipedia.org/w/index.php?title=Square_roo
        t_of_2&oldid=1175903907

9.3   "BBP formula for e"
        Mathematics Stack Exchange
        Number Theory Pages
        JacobTDC, edited Feb 8, 2018 at 16:03
        JacobTDC writes "Please note that this is not a proper
        BBP formula, but can still be used in the digit-
        extraction algorithm."
        https://math.stackexchange.com/questions/1817064/bb
p-formula-for-e

Chapter Ten

10.1 "The ichnology of the Winterhouse Formation"
By
©Robyn Rebecca Reynolds, B.Sc. (Hons)
A thesis submitted to the
School of Graduate Studies
in partial fulfilment of the requirements
for the degree of
Master of Science
Department of Earth Sciences
Memorial University of Newfoundland
St Johns, Newfoundland
October 2015
pp218

10.2 ICHNOLOGY OF VISEAN SANDSTONES
IN NORTHWESTERN IRELAND:
A STUDY OF TRACE FOSSILS IN THEIR
PALAEOECOLOGICAL
AND SEDIMENTOLOGICAL CONTEXT
Thesis submitted for the degree of
Doctor of Philosophy
in the Faculty of Science, University of London
by
AVINASH CHANDRA
B.Sc.(Hons.), M.Sc.
August 1974
pp323

10.3 BIOTURBATION AND TRACE FOSSILS
IN DEEP SEA
SEDIMENTS OF THE WALVIS RIDGE,
SOUTHEASTERN ATLANTIC, LEG 741
Dieter K. Putterer,
Geologisch-Palaontologisches Institut der
Universitat Kiel,D-2300 Kiel,
Federal Republic of Germany
http://deepseadrilling.org/74/volume/dsdp74_12.pdf
pp13

10.4   Попова Л. В.
        Paleodiction – іхнотаксон чи фізичне явище?
        // Вісник Київського університету. Серія Геологія.
        – 2009. – Вип. 47. – С. 8- 10.
        January 2009
        Author:
        Lilia Popova
        National Academy of Sciences of Ukraine
        I. I. Schmalhausen Institute of Zoology
        In Cyrillic Script and Russian Language
        with an Abstract in Ukranian

10.5   Wikipedia contributors. (2023, August 18).
        Froude number.
        In *Wikipedia, The Free Encyclopedia*.
        Retrieved 09:28, October 18, 2023, from
        https://en.wikipedia.org/w/index.php?title=Froude_nu
        mber&oldid=1171005585

10.6   *Paleodictyon*, probably *P. imperfectum*, in the
        flysch sandstone at Punta de San García.
        Campo de Gibraltar complexes,
        Algeciras (Oligocene-Miocene) unit,
        Algeciras, Andalusia, Spain.
        Photograph by Falconaumanni
        24 December 2014, 14:22:25
        in
        Wikipedia contributors. (2023, June 7).
        Paleodictyon.
        In *Wikipedia, The Free Encyclopedia*.
        Retrieved 17:29, October 17, 2023, from
        https://en.wikipedia.org/w/index.php?title=Paleodicty
        on&oldid=1158965617

10.7   Wikipedia contributors. (2023, July 18).
Rayleigh–Bénard convection.
In *Wikipedia, The Free Encyclopedia*.
Retrieved 15:29, October 15, 2023, from
https://en.wikipedia.org/w/index.php?title=Rayleigh%
E2%80%93B%C3%A9nard_convection&oldid=1165987910

10.8   "The formation of columnar joints produced by
cooling in basalt at Staffa, Scotland"
JC Phillips, MCS Humphreys,
KA Daniels, RJ Brown, F Witham
Bulletin of Volcanology (2013) 75:715
DOI 10.1007/s00445-013-0715-4
Springer Verlag of Berlin and Heidelberg
22 November 2010
pp17
https://www.geology.cwu.edu/facstaff/nick/g351/2013_Philli
ps_Columns.pdf

10.9   "Early Cambrian Microburrow Nests and the
Origin of Parenting Skills"
(Conference Paper)
0830 4 November 2012
Charlotte, North Carolina, USA
The Geological Society of America
Annual Meeting and Exposition
MCMENAMIN, Mark A.S., Geology and Geography,
Mount Holyoke College, South Hadley,
MA 01075, mmcmenam@mtholyoke.edu
Paper No. 210-3
https://d1wqtxts1xzle7.cloudfront.net/34691240/CambrianMicroburr
ows-libre.pdf?1410374471=&response-content-
disposition=inline%3B+filename%3DEarly_Cambrian_microburrow
_nests_and_the.pdf&Expires=1697456106&Signature=OiKwfXxpsy
qc1euHip5BBIYmAEHIE0hsqyXGB6U8XWZApEJ9H6AkxVjNPqb
8gpdrgBnKNSjJbBhxWroT8nkSbYL7hUG5MXMdrKCNUW1wo1tj
eBlGBDNuEwguEMolZCZmP3AaFwkOoMLNGayhs46i~BS~1sUL
pOYDs9tZ0QlfC9hhjOHqQoJrQoqL0crEoUxwCGta3XQisvF6GmO
DXwcKhNV1ZL7HYE4gxa7z0ShuAiKh2098MPY2HU3V7BhkV8j
CAlpT0MIXofqkXssYmnddzrEu6HWdREQrLGbpbOGKhCXexJe8
Z~yTF5hK1prL9x1S1C95DdpG7X6JbCBbNhTqZw__&Key-Pair-
Id=APKAJLOHF5GGSLRBV4ZA

with downloadable PowerPoint File Handout
MicroburrowNests3.pptx (6.9 MB)

<u>Appendix C</u>

C1   Wikipedia contributors. (2023, August 18).
Froude number.
In *Wikipedia, The Free Encyclopedia*.
Retrieved 13:50, October 21, 2023, from
https://en.wikipedia.org/w/index.php?title=Froude_number&oldid=1171005585

C2   "The formation of columnar joints produced by
cooling in basalt at Staffa, Scotland"
J. C. Phillips & M. C. S. Humphreys & K. A. Daniels
& R. J. Brown & F. Witham
Received: 22 November 2010
Accepted: 27 February 2013
Springer-Verlag Berlin Heidelberg 2013
Bull Volcanol (2013) 75:715
DOI 10.1007/s00445-013-0715-4
pp17

C3   Wikipedia contributors. (2023, September 22).
Reynolds number.
In *Wikipedia, The Free Encyclopedia*.
Retrieved 13:48, October 21, 2023, from
https://en.wikipedia.org/w/index.php?title=Reynolds_number&oldid=1176628646

C4   "Convective heat transfer in a thermal chimney
for freshwater production in geothermal
total flow systems"
Wenguang Li, Guopeng Yu, Zhibin Yu
13 January 2023
Applied Thermal Engineering 230 (2023) 120848
www.elsevier.com/locate/apthermeng
https://www.sciencedirect.com/science/article/pii/S1359431123008773?via%3Dihub
pp26

APPENDIX A
SOME CONVENIENT FUNCTIONS

The Percentage Specific Defect

The Specific Defect is a metric of the relative difference between two numbers, preserving the sign of the result. It is especially useful for comparing the difference between x, the Fiducial Number computed by a definitional or other trusted source, and y, the provisional value under consideration. It therefore compares for a discrepancy.

Our habit is for clarity to compute the Percentage Specific Defect, PSD(x,y) defined by:-

$$PSD(x, y) = 100 \left(\frac{x - y}{x} \right)$$

Equation A1

Exactitude (within computational tolerance) is PSD(x,y) = 0.

The Double Factorial Function DFAC

DFAC is defined by:-

$$DFAC(n) = \prod_{k=0}^{ceil\left(\frac{n}{2}\right)-1} (n - 2k)$$

Equation A2

For example, DFAC(8) = 8×6×4×2 = 384

Table A1 presents the first few examples of DFAC for both simple products and for the EXCEL® intrinsic function FACTDOUBLE(n).

Equation A2 may be elaborated as:-

$$n!! = DFAC(n) = \prod_{k=0}^{ceil\left(\frac{n}{2}\right)-1} (n - 2k) = n \times (n - 2) \times (n - 4) \dots \times 1$$

Equation A3

For example, in the case of odd n, n = 9 for sake of argument, the full series is:-

$$9!! = DFAC(9) = \prod_{k=0}^{ceil\left(\frac{9}{2}\right)-1} (9 - 2k) = 9 \times 7 \times 5 \times 3 \times 1 = 945$$

Equation A4

and for even n, n = ten say:-

$$10!! = DFAC(10)$$

$$= \prod_{k=0}^{ceil\left(\frac{10}{2}\right)-1} (10 - 2k) = 10 \times 8 \times 6 \times 4 \times 2 \times 1$$

$$= 3840$$

Equation A5

In terms of both MathCad® Express® and EXCEL® the computation of double factorials is always problematical, largely because of the very steep increase of n!! with n.

At n = 29, the fifteen-digit n!! value is 6190283353629380, and the MathCad capacity is reached. Similar exhaustions restrict EXCEL®, and indeed its intrinsic function FACTDOUBLE().

Resolved functional forms such as:-

$$DFACODD(n) = \prod_{i=1}^{\frac{n-2}{2}+1} (2i + 1)$$

Equation A6a

$$DFACEVEN(n) = \prod_{i=1}^{\frac{n-2}{2}+1} (2i)$$

Equation A6b

unfortunately offer no mitigation.

I also studied whether the user-defined MathCad® Express® function:-

$$\Gamma(z) = \int_0^{96} t^{z-1} e^{-t}. dt$$

Equation A7

would be helpful in association with the expression:-

$$n!! = round\left[\sqrt{\frac{2}{\pi}} . 2^{\frac{z}{2}} . \Gamma\left(1 + \frac{z}{2}\right)\right] = round[dg_k]$$

Equation A8

dg_k is inaccurate in the form in which I have expressed it and Equation A8 as a whole is unreliable for even n

The Finite Limits of Computational Machinery

Most practical modern computers and calculators base their electronic processing upon arrays of bistable-multivibrators, which are themselves coupled triode circuits. These circuits are printed photolithographically at the rate of thousands per square centimeter, so that individual triodes are microscopic, but data handling capacities finitely prodigious.

A bistable-multivibrator has the property that it can switch from a "zero" sub-circuit to a "one" sub-circuit when receiving a stimulus pulse of electricity: And switch back again when experiencing a second prompt. These quasi-stable states maintain as long as electric power is supplied.

Such physical arrays are always finite, even though we attempt mathematically to model the infinite.

My Arithmetic Logic Unit (ALU) is notionally a row of sixty-four multivibrators which represent a 64-digit base-2 (i.e. binary) number. Binary numbers can always be converted to denary numbers and vice-versa, whether by hand or by machine.

Any computed number has to fit this electronic scheme, but there are many hardware and programming tricks that can be used to boost the length and precision of numbers. Such tricks are expensive in time and complexity, and are usually justified only in critical applications. Most of us find fifteen-digit denary precision more than adequate, and fifteen digits denary can neatly be represented by a 64-digit hardware register as we may mathematically demonstrate:-

Allow that n is the Array Width in (Binary) Digits (i.e. 64 for a normal microcomputer in 2023AD); x is the notional Denary Exponent of the Limit Number (say 128); s_1 is the Sign Bit for the Mantissa; s_2 is the Sign Bit for the Exponent; p is the Exponent Bit Width; and m is the Mantissa Bit Width.

Then the Mantissa Bit Width is:-

$$m = n - s_1 - s_2 - p = 64 - 1 - 1 - 7 = 55$$
Equation A9

because:-

$$p = 128 = 2^7$$
Equation A10

and Binary and Denary are interconvertible by:-

$$2^m = 10^x$$
Equation A11

Taking Base-10 logarithms of each side of Equation A11 we have:-

$$x = m.\log_{10} 2$$
Equation A12

so that:-

$$10^x = 3.60287970189639 \times 10^{16}$$
Equation A13

Taking Base-10 logarithms:-

$$\log_{10}(10^x) = x = 16.556649761519$$
Equation A14

Therefore a 64-bit register has a theoretical denary capacity in excess of 16 denary digits, so fifteen digits can be handled by a 64-bit register with some space to spare.

If you have a 128-bit processor you can probably achieve 35 decimal digits without resorting to software double-precision or other stratagems.

The Binomial Expansion Function BINOM(n,k)

The n,k Coefficient of a Binomial Expansion is given by:-

$$BINOM(n,k) = \frac{n!}{k!\,(n-k)!}$$
Equation A15

where the screech symbol represents the Factorial Function:-

$$n! = \prod_{k=1}^{n} k$$
Equation A16

For example:-

$$5! = \prod_{k=1}^{5} k = 1 \times 2 \times 3 \times 4 \times 5 = 120$$
Equation A17

The Weierstrass Integral (1841)

The Weierstrass Integral is the theoretical definition of π the Ludolphine Constant, i.e. the Ratio of the Circumference of a Circle to its Diameter.

The Weierstrass Integral is:-

$$\pi_{Weierstrass} = \int_{-1}^{+1} \frac{1}{\sqrt{1-x^2}}\,dx = 3.14159265349741$$
Equation A18

The Weierstrass Integral is subject to numerical error in real-life computations.

MathCad® Express® quoted π to be 3.14159265358979, and so the PSD attaching to the Weierstrass Integral as elaborated by Express® was:-

$$PSD\big(\pi_{fido}, \pi_{Weierstrass}\big) = 0.000000002940571$$
Equation A19

n	Reference DFAC	FACTDOUBLE(n)
0		1
1	1	1
2	2	2
3	3	3
4	8	8
5	15	15
6	48	48
7	105	105
8	384	384
9	945	945
10	3840	3840
11	10395	10395
12	46080	46080
13		135135
14		645120
15		2027025
16		10321920
17		34459425
18		185794560
19		654729075
20		3715891200
21		13749310575
22		81749606400
23		3.16234E+11
24		1.96199E+12
25		7.90585E+12
26		5.10118E+13
27		2.13458E+14
28		1.42833E+15
29		6.19028E+15
30		4.28499E+16

Table A1
Double Factorials
Simple Product DFAC to n = 12 and
EXCEL® Intrinsic FACTDOUBLE to n = 30

The general algebraic polynomial P(n) is defined as:-

$$P(n) = \sum_{k=0}^{n} c_k . x^k$$

Equation A20

where n is the Degree of the Equation (1 for Linear; 2 for Quadratic; 3 for Cubic, etcetera); c_k are Coefficients (positive or negative); and k is the Serial of the Summative Term.

Our chosen Analytical Test Function is a simple second-degree algebraic polynomial (i.e. a quadratic equation):-

$$P(2) = \sum_{k=0}^{2} c_k . x^k = 2.333 + \frac{1}{5} \times x + 14 \times x^2$$

Equation A21

where $c_0 = 2.333$, $c_1 = 1/5$ and $c_2 = 14$

The Analytic Integral $I_{P(2)}$ of an Algebraic Polynomial is given by:-

$$I_{P(n)} = C + \sum_{k=0}^{n} c_k . \frac{1}{k+1} . x^{(k+1)}$$

Equation A22

where C is a Constant of Integration, which in the case of Definite Integrals cancels out and may be ignored.

A Definite Integral is defined between a Lower Bound, lb, of the independent variable x, and an Upper Bound ub.

For example, in our case:-

$$I_{Def} = \int_{lb}^{ub} \left(14x^2 + \frac{1}{5}x + 2.333\right) dx = \int_{0}^{8} \left(14x^2 + \frac{1}{5}x + 2.333\right)$$

Equation A23

because we will determine the Integral between x = 0 and x = 8

Therefore:-

$$I_{Def} = \int_{lb}^{ub} \left[14(x)^2 + \frac{1}{5}(x) + 2.333\right] dx$$

$$= {}_{0}\left[14 \times \frac{1}{3}x^3 + \frac{1}{5} \times \frac{1}{2} \times x^2 + 2.333 \times x\right]^{8}$$

$$= {}_{0}\left[14 \times \frac{1}{3}(ub - lb)^3 + \frac{1}{5} \times \frac{1}{2} \times (ub - lb)^2 + 2.333 \times (ub - lb)\right]^{8}$$

$$= 14 \times \frac{1}{3}8^3 + \frac{1}{5} \times \frac{1}{2} \times 8^2 + 2.333 \times 8$$

Equation A24

The numerical value of I_{Def} for our example is 2414.39733333333

In Functional Terms, we may define for the general quadratic:-

$$FUN(x, \alpha, \beta, \gamma) = \alpha.x^2 + \beta.x + \gamma$$

Equation A25a

$$FUNI(x, \alpha, \beta, \gamma) = \frac{\alpha}{3}.x^3 + \frac{\beta}{2}.x^2 + \frac{\gamma}{1}.x^1$$

Equation A25b

$$FUNIDEF(lb, ub, \alpha, \beta, \gamma)$$
$$= \frac{\alpha}{3}.(ub - lb)^3 + \frac{\beta}{2}.(ub - lb)^2 + \gamma.(ub - lb)$$

Equation A25c

so that:-

$$FUN(2, \alpha, \beta, \gamma) = 58.733$$
Equation A26a

$$FUNI(2, \alpha, \beta, \gamma) = 42.3993333333333$$
Equation A26b

$$FUNIDEF(lb, ub, \alpha, \beta, \gamma) = 2414.39733333333$$
Equation A26c

The Lerch Transcendent Function (1887)

The Lerch Transcendent is defined as:-

$$\Upsilon(z, s, \alpha) = \sum_{n=0}^{48} \frac{z^n}{(n+\alpha)^s}$$
Equation A27

The upper limit of the summation is flexible.
For example, given the current RHS, $\Upsilon(1,1,1) = 4.47920533832942$

Newton-Cotes Formulae: The Extended Simpson's Rule

The Extended Simpson's Rule is a method of numerical integration which attempts in a virtual way to fit a series of parabolic arcs to a curvilinear line in order to define an approximate quadratic fit. This is tantamount to approximating every three points to the parameters of a parabola as a scheme of integration yielding the area under the whole curve.

Simpson's Rule is one of a family of Newton-Coates Formulae that treat an {x,y} point-series as an algebraic polynomial. There are only eight credible algebraic polynomial structures: Linear, Quadratic, Cubic, Quartic, Quintic, Sextic, Septic and Octic. There are an infinite number of algebraic structures "in Nature".

Therefore almost all circumstances in which we apply Newton-Cotes are theoretically-inappropriate. But nevertheless, Numerical Integration schemes of all sorts are unbelievably useful, especially when an analytical integral does not exist, or at least has not yet been discovered.

We just need to be careful and realistic in what we demand of these tools.

In particular, few of the laws of science are more than quartic in general scope or pattern, and attempts to apply Newton-Cotes principles to schemes that exceed the eighth degree invoke numerical instabilities in machinery that cause corrupt outputs.

The General Extended Simpson's Rule Formula is given by:-

$$SIMP(m, lb, ub, h, x)$$
$$= \frac{h}{3}\left\{FUN(lb, \alpha, \beta, \gamma)\right.$$
$$+ \sum_{k=1}^{m-1}[3 - (-1)^k] \cdot FUN(k \cdot h, \alpha, \beta, \gamma)$$
$$\left. + FUN(ub, \alpha, \beta, \gamma)\right\}$$

Equation A28

where m is the Number of Intervals, one less than the Number of Data Points, which includes $k = 0$. The Number of Intervals must be even. The minimum number of points m+1 is three.

lb and ub are respectively the Lower Bound and the Upper Bound x of the implied Definite Integral represented by the RHS.

$\alpha, \beta, \gamma \equiv c_2, c_1, c_0 \equiv 14, 0.2, 2.333$ the Coefficients of the Equation A21 Trial Polynomial. These parameters could be any positive or negative real number.

h is the Series Interval Width given by:-

$$h = \frac{ub - lb}{m}$$

Equation A29

For Closed-Form Newton-Cotes methods the Interval Width must be constant and hence h is constant for any particular job.

Because we are using an exact quadratic polynomial as our test subject it is exactly fitted by a quadratic curve defined by three points.

Accordingly, any scheme where m exceeds three is massively redundant in our example.

The numerical values of SIMP(2,lb,ub,h,x), SIMP(16,lb,ub,h,x), SIMP(32,lb,ub,h,x), SIMP(64,lb,ub,h,x), and SIMP(128,lb,ub,h,x), are all 2414.397333 within limits of accuracy.

Or to put it another way:-

$$I_{Def} = \left(\alpha \cdot \frac{1}{3} \cdot ub^3 + \beta \cdot \frac{1}{2} \cdot ub^2 + \gamma \cdot \frac{1}{1} \cdot ub^1 \right)$$
$$- \left(\alpha \cdot \frac{1}{3} \cdot lb^3 + \beta \cdot \frac{1}{2} \cdot lb^2 + \gamma \cdot \frac{1}{1} \cdot lb^1 \right)$$
$$= 2414.39733333333$$

Equation A30

whilst:-

$$SIMP(2, lb, ub, h, x) = 2414.39733333333$$

Equation A31

In my experience, the Extended Simpson's Rule is simultaneously the most compact, accurate, economical and foolproof of all the Newton-Cotes Integrators for nearly all applications.

But others may be useful in specialized situations.

In terms of Abramowitz and Stegun, the Extended Simpson's Rule is A&S 25.4.6

Table A2 shows the 16-inteval term series and results for our test equation whilst Table A3 shows the exact same output for the necessary and sufficient 2-interval elaboration.

Newton-Cotes Formulae: Rule 18

In terms of Abramowitz and Stegun, the octic-fitting Rule 18 is A&S 25.4.18

Rule 18 is a very powerful and relatively stable numerical integration formula, but like all mathematical objects has to be used cautiously and skillfully, employing all of your experience and scientific insight.

Rule 18 is capable of extension as are all Newton-Coates Formulae: You just double the co-efficient at the interior series contacts. Whether this is worth doing with Rule 18 is another matter...

We will stick to m = 8.

Rule 18 is massively overpowered for a quadratic integration, but for demonstrational purposes we will study the 8-interval solutions case and the way in which the nine coefficients could be lapped for the 16-interval case: Rule 18 can be extended to m = 8n, where n is an arbitrary integer. As previously remarked, whether such extension is justified is a matter for your judgment.

Table A5 (non-consecutive) presents the lists of coefficients for the 8 and 16-interval methods. Note that c_8 is doubled for the 16-interval exercise.

k	c_k	c_k
0	989	989
1	5888	5888
2	-928	-928
3	10496	10496
4	-4540	-4540
5	10496	10496
6	-928	-928
7	5888	5888
8	989	1978
9		5888
10		-928
11		10496
12		-4540
13		10496
14		-928
15		5888
16		989

Table A5
Rule 18 Coefficients

For the test coefficients α = 14, β = 0.2, and γ = 2.333 the quadratic is defined in the following terms:-

$$FUN(x, \alpha, \beta, \gamma) = \alpha x^2 + \beta x + \gamma$$
Equation A32

which may be elaborated as:-

$$FUN(2, \alpha, \beta, \gamma) = 58.733$$
Equation A33

It follows that the Rule 18 summation FUN18(h,α,β,γ) may be computed in the MathCad® Express® idiom as:-

$$FUN18(h, \alpha, \beta, \gamma) = \sum_{k=0}^{8} c_k . FUN(k, 14, .2, 2.333) = 8556020.55$$

Equation A34

from which the actual Rule 18 integration may be defined as:-

$$RULE18(h, \alpha, \beta, \gamma) = \frac{4}{14175} . h . FUN18(h, \alpha, \beta, \gamma)$$

Equation A35

so that:-

$$RULE18(h, \alpha, \beta, \gamma) = RULE18(1,14,0.2,2.333)$$
$$= 2414.39733333333$$

Equation A36

Alternatively, a two-step approach may be convenient in which:-

$$Soln = RULE18(h, \alpha, \beta, \gamma)$$

Equation A37a

$$Soln = 2414.39733333333$$

Equation A37b

Number of Intervals	m	16	
Integral Lower Bound	lb	0	
Integral Upper Bound	ub	8	
Interval Width	h	0.5	

ANALYTIC TEST FUNCTION

	x	2
	α	14
	β	0.2
	χ	2.333
Function Value	F	58.733
Analytic Integral Of Function	I_{an}	42.39933333
Integral at Lower Bound	I_{lb}	0
Integral at Upper Bound	I_{ub}	2414.397333
Definate Integral of Function	I_{def}	2414.397333
Simpson Multiplier	h/3	0.166666667
Simpson Sum of Terms	Σ_{simp}	14486.384
Simpson Integral	I_{simp}	2414.397333
PSD(I_{def},I_{simp})		1.88348E-14

k	k*h	c_k	Function f(k)	Term
0	0	1	2.333	2.333
1	0.5	4	5.933	23.732
2	1	2	16.533	33.066
3	1.5	4	34.133	136.532
4	2	2	58.733	117.466
5	2.5	4	90.333	361.332
6	3	2	128.933	257.866
7	3.5	4	174.533	698.132
8	4	2	227.133	454.266
9	4.5	4	286.733	1146.932
10	5	2	353.333	706.666
11	5.5	4	426.933	1707.732
12	6	2	507.533	1015.066
13	6.5	4	595.133	2380.532
14	7	2	689.733	1379.466
15	7.5	4	791.333	3165.332
16	8	1	899.933	899.933

Table A2
A 16-interval Extended Simpson's Rule Elaboration for the
Quadratic Equation $14x^2+0.2x+2.333$

Number of Intervals	m	2					
Integral Lower Bound	lb	0					
Integral Upper Bound	ub	8					
Interval Width	h	4					
ANALYTIC TEST FUNCTION							
	x	2					
	α	14					
	β	0.2					
	χ	2.333					
Function Value	F	58.733					
Analytic Integral Of Function	I_{an}	42.39933333					
Integral at Lower Bound	I_{lb}	0					
Integral at Upper Bound	I_{ub}	2414.397333					
Definate Integral of Function	I_{def}	2414.397333					
Simpson Multiplier	h/3	1.333333333					
Simpson Sum of Terms	Σ_{simp}	1810.798					
Simpson Integral	I_{simp}	2414.397333					
PSD(I_{def},I_{simp})		1.88348E-14					

k	k*h	c_k	Function f(k)	Term
0	0	1	2.333	2.333
1	4	4	227.133	908.532
2	8	1	899.933	899.933

Table A3
A 2-interval Extended Simpson's Rule Elaboration
Necessary and Sufficient to the Application for the
Quadratic Equation $14x^2+0.2x+2.333$

Number of Intervals	m	8	
Integral Lower Bound	lb	0	
Integral Upper Bound	ub	8	
Interval Width	h	1	

ANALYTIC TEST FUNCTION

	x	2	
	α	14	
	β	0.2	
	χ	2.333	

Function Value	F	58.733
Analytic Integral Of Function	I_{an}	42.39933333

Integral at Lower Bound	I_{lb}	0
Integral at Upper Bound	I_{ub}	2414.397333
Definate Integral of Function	I_{def}	2414.397333

Rule 18 Multiplier	4h/14175	0.000282187
Rule 18 Sum of Terms	Σ_{rule18}	8556020.55
Rule 18 Integral	I_{rule18}	2414.397333

PSD(I_{def},I_{rule18})		3.76696E-14

k	k*h	c_k	Function f(k)	Term
0	0	989	2.333	2307.337
1	1	5888	16.533	97346.3
2	2	-928	58.733	-54504.2
3	3	10496	128.933	1353281
4	4	-4540	227.133	-1031184
5	5	10496	353.333	3708583
6	6	-928	507.533	-470991
7	7	5888	689.733	4061148
8	8	989	899.933	890033.7

Table A4
An 8-interval Rule 18 Elaboration
Redundantly-Sufficient to the Application for the
Quadratic Equation $14x^2+0.2x+2.333$

APPENDIX B
REFERENCE VALUES

For corroborative purposes we may present twenty-one digit numeric values of our four famous numbers as:-

π	3.14159 26535 89793 23846
e	2.71828 18284 59045 23536
Φ	1.61803 39887 49894 84820
Ψ	1.41421 35623 73095 04880

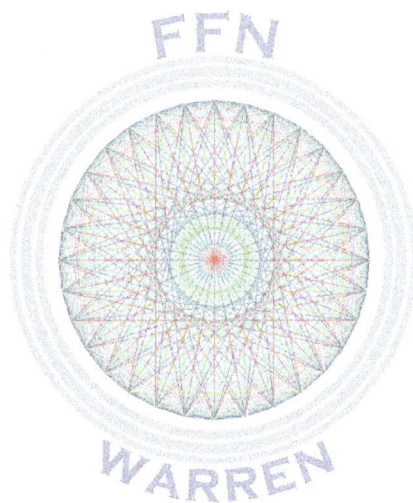

APPENDIX C
THE CONVECTION OF BASALTIC MAGMA

Convection is a very complex process, and highly mathematical. Our discussion will be very elementary and focus on very basic perplexities. If we are lucky or skillful we shall have the opportunity to re-visit some old friends, not only the dead geniuses of the past, but also a brace or two of our Famous Numbers.

Since before the dawn of our Industrial Enlightenment men have striven to understand the flow of fluids, the movement of blood and oil, of wine and of water. Or, as a wise king once observed, the way of a ship on the sea.

When fluid moves there is drag. Drag is a force that dissipates energy, which costs money, and energy was always expensive in Britain. Drag impedes progress (literally) and may well prevent a powered vessel reaching America, or render flight impossible.

When, in the eighteenth-century my country sought to trade its way to survival it invested in forests of smokestacks, each of which was a necessary enemy to production. All were convective, but many were difficult to work safely or profitably. Design was controversial, and all sorts of features: Oversailors, dampers, linings, even height were often ineffectual.

So what is convection? Convection is the tendency of a lighter substance, even hot air, buoyantly to rise amidst denser material. When most fluids, and certainly air, are heated they expand, and mass-for-mass their density lightens and they rise. In semi-closed systems like chimneys or volcanic magma chambers, colder material displaced descends to replace the hot material according to the Laws of Continuity.

Another source of drag is viscosity. Viscosity, which comes in Kinematic and Dynamic flavors, is a tendency (pedantically, it is incorrect to call it a "force") to retard the motion of a fluid against a surface, or even against itself. Viscosity contributes to drag. Viscosity mirrors the "thickness" of a fluid. Pitch, treacle and asphalt are viscous: Water, helium and gasoline are not.

In the design and construction of chimneys we earnestly hope that fluid flow will strictly be in one direction!

Work on the design of chimneys continues in 2023AD, at least in Glasgow[C4], where the construction of tall chimneys originated in attempts to disperse the muriatic (hydrochloric) acid gas byproduct of LeBlanc alkali production that otherwise was observed to poison the population and eat the mortar from wet brick walls.

Laminar and Turbulent Flow

A clear distinction can be made between laminar and turbulent flow. Turbulent is much less energy-efficient than laminar flow.

Laminar flow involves the molecules of liquid sliding parallel to each other or to the surface along which they travel. Turbulent flow involves churning and roiling of the fluid in several directions, though the gross flow may tend in one direction.

I have schematically illustrated the difference between the two flow types in Figure C1a and Figure C1b.

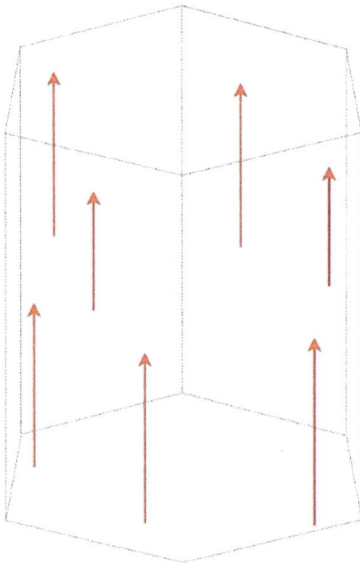

Figure C1a
Laminar Flow

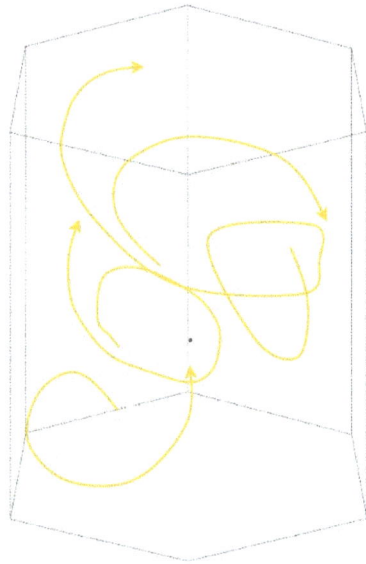

Figure C1b
Turbulent Flow

The setting of the two flow types is a sub-hexagonally sectioned hexagonal chimney or vertical conduit, an idealisation of a solidifying basaltic lava column that we may later discuss.

The Froude Number[C1,C2] and the Reynold's Number[C3,C4] are two dimensionless continuum-mechanical parameters of hydraulic physics. As observed elsewhere, the Froude Number is essentially the ratio of kinetic to potential energy in a mobile system but in the context of fluid physics it may be defined:-

$$Fr = \frac{u}{\sqrt{gL}}$$

Equation C1

and is therefore, in context, the ratio of inertial to gravitational forces.

Fr is the Froude Number; u is the gross Fluid Velocity; g is the (Standard) Acceleration dur to Gravity; and L is a "Characteristic Length".

Identifying the "Characteristic Length" of the system is a problematic matter. It is easy to assume that the "Characteristic Length" of a chimney is its Height, but it is more sensible to adopt the internal diameter (whatever that is) as L. In the case of a ship, naval architects usually prefer to adopt the ship's length as L, but is that keel length, waterline line length,...

When some people have attempted to compute the Froude Number for the magma columns of forming basaltic colonnades they have made the same mistake as myself and used Column Height as L. It is sounder to use the diameter of a magma column as L. The big problem is deciding what qualifies as the "diameter" of a somewhat irregular natural polygon.

Conveniently, laminar flow transitions to a turbulent state when Fr is unity. Therefore, we can readily compute the long-stilled velocity of a lava emergence by transposing Equation C1 *if we can assume that Fr was one at the time*:-

$$u = (Fr).\sqrt{gL}$$

Equation C2

More realistically, if the rock fabric supports laminar flow we may compute the local *maximum* magma velocity, given tenable values of g and L.

There may be good geological evidence, fossilised in the rock fabric, to support laminar convection or even laminar extrusion.

Another vagary is the choice of "diameter". What is the "diameter" of an ideal hexagon? What is the "diameter" of an irregular hexagon, or indeed pentagon or heptagon as seen in a geological locality?

Figure C2 illustrates a plausible four of the infinite possible choices:-

(a) The Diameter of the Inscribed Circle
(b) The Diameter of the Equal-Area Circle
(c) The Diameter of the Circle of Equal Perimeter
(d) The Diameter of the Exscribed (escribed) Circle

The Red Circle is the Incircle inscribed within the regular hexagon in black. The Blue Circle is the Circumcircle, or if you prefer Excircle which touches each hexagon vertex from the outside. The Green Circle is the circle whose area is equal to the area of the hexagon. Fourthly, the Purple Circle is the circle whose perimeter is equal to the perimeter of the hexagon (i.e. 6t where t is the length of a Regular Hexagon Side).

Given that the hexagon side length, t, is unity, the respective Circle "Characteristic Diameters" are:-

(a) $D_{Incircle} = 2r = 1.73205080756888$

(b) $D_{EqualAreas} = 2.\rho_A.t = 1.8187834869854$

(c) $D_{EqualPerimeters} = 2.\rho_P.t = 1.90985931710274$

(d) $D_{Excircle} = 2R = 2$

What a quandary! Will the true "characteristic diameter" please stand up! In science, as in politics and religion the answers are by no means as clear-cut as most people think. Even the questions are often obscure.

My own prejudices are based upon the concepts of "wetted perimeter" and its velocity-produced "wetted surface", because I think that these are the only geometric elements upon which viscous drag logically may act.

Accordingly I favor $D_{EqualPerimeters}$ as "characteristic length" L.

A second dilemma: What is the "velocity" of a current of viscous fluid rising against drag? Figure C3 displays a sub-parabolic velocity profile for a fluid progressing against the friction of the walls of its conduit. The fastest motion is in the center of the conduit. The speed of the fluid is theoretically zero at the wall. The "uniform velocity" is proportional to the height of the purple rectangle which represents a three-dimensional integral of the paraboloid velocity-cone. That is to say that the fine green shaded area is equal to the fine blue shaded vertical area given symmetry.

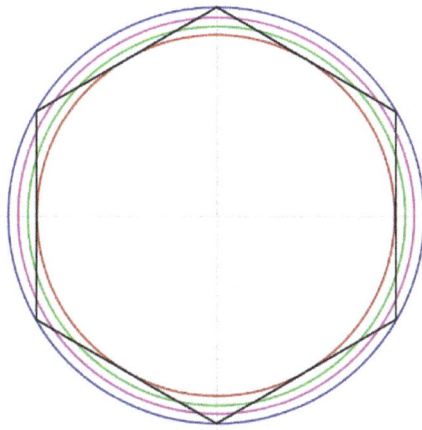

Figure C2
Alternative Characteristic Diameter Circles
for a Regular Hexagon

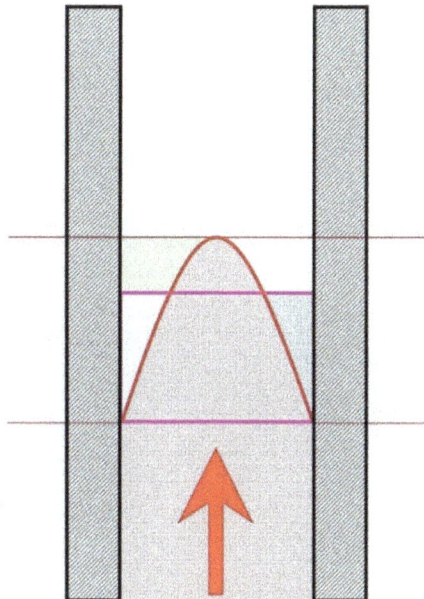

Figure C3
Sketch Diagram of Velocity Cone
for Basaltic Magma rising through a Rock Fissure
as a Viscous Column

The Polygonal Basaltic Columns of Staffa

In 2013AD, Phillips Et Al[C2] reported upon their valuable studies of the British Tertiary Volcanic Province basaltic formations at the Isle of Staffa in the Inner Hebrides islands of Scotland (NM 32460 35549: 56° 26′ 10″ N, 6° 20′ 31″ W).

The word "Staffa" may derive from the Old Norse words stafr and öy, meaning "Stave Island" or "Staff Island", a term easy to understand from the seaboard southern sight of the island as shown in my photograph Figure C4. Given that the sea horizon is at approximate relative height 0.55 Fingal's Cave is the black hollow at (0.73. 0.55) and Am Buachaille the low conical stack at (0.86, 0.55).

Am Buachaille, often translated from the Gaelic as "The Shepherd" is more appropriately rendered "The Sentinel" in the context of the numerous Highland sea-stacks that bear this appellation, those that can be fancied to guard beaches or little harbors.

Photo: James R Warren

Figure C4
Leaving Staffa from the South

With regard to basaltic column geometries Phillips Et Al measured at Points 1, 2, 3, 5, 6 ad 8 as located upon their Figure C5 map.

I have abstracted these data to Table C1. I have rendered their CGS measurements to SI.

Phillips Et Al determined the arithmetic mean Column Diameter to be 0.881 meters and the mean column Side Length to be 0.414833 meters, whilst the Column Top Area was 0.431283 meters^2 on average and the mean polygonal top Internal Angle was 2.076942 radians.

The normative parameters for a regular hexagon, where Side Length t is unity, are:-

The Normative Column (Hexagon) Diameter, D_{norm}

$$D_{norm} = \frac{\sqrt{3}}{2} \times t = 0.866025$$

Equation C3

The Normative Top Area, A_{norm}

$$A_{norm} = \frac{3\sqrt{3}}{8} \cdot D_\mu^2 = 0.504131358$$

Equation C4

The Normative Mean Internal Angle, α_μ

$$\alpha_\mu = \frac{2\pi}{3} = 2.094395102$$

Equation C5

Hence defining the Percentage Specific Defect in our usual way as:-

$$PSD(DataMean, Norm) = 100 \left(\frac{DataMean - Norm}{Norm} \right)$$

Equation C6

we compute:-

$$PSD(D_\mu D_{norm}) = 1.699727153$$
Equation C7a

$$PSD(A_\mu A_{norm}) = -16.89098989$$
Equation C7b

$$PSD(\alpha_\mu, \alpha_{norm}) = -0.840336134$$
Equation C7c

I ascribe the Diameter Estimation PSD, not significantly different to 1.5%, to field measurement error, always difficult in hostile environments like Scotland, notwithstanding my idyllic-looking photograph.

Similarly the goniometry displays less than one percent error. There are three *independent* internal angles in a hexagon, reflecting the fact that the total of the internal angles of a polygon is $\pi(N-2)$ radians. The very large error in Area Estimation is almost certainly due to the known existence of both irregular hexagons, and many non-hexagons, among the column tops at Staffa.

The Phillips team define the Diameter D of a basalt column top to be the Maximal Intervertex Distance. In terms of the ideal hexagon this is equivalent to $2R \equiv 2t$ where R is the Radius of the Excribed Circle and t is the Hexagon Side Length.

When standardised for $t = 1$, the average D is 2.123744476 which implies a PSD($D_{ideal}, D_{measured}$) of -6.1872. I find it difficult to resolve which part of this discrepancy is due to field error, and which part due to the fact that we are not dealing with mathematically-perfect hexagons.

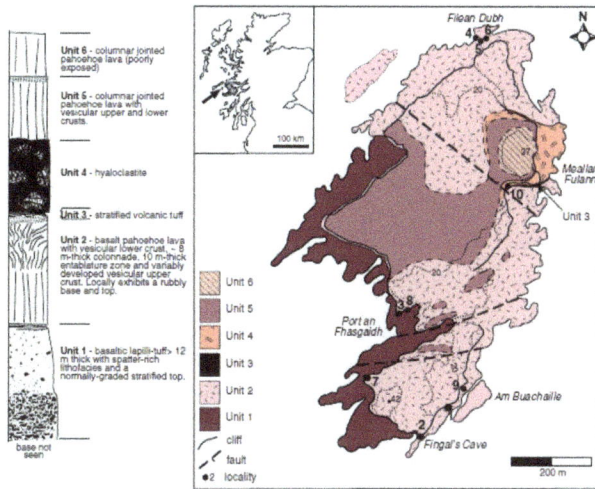

Fig. 3 Geological map (*right*) and stratigraphic section (*left*) for Staffa. *Inset* shows location of the Isle of Mull (*shaded*) with *arrow* to indicate the location of the island of Staffa. Localities studied are marked with *large dots* and a *locality number* (*small dots* indicate topographic spot heights)

Drawings: Phillips Et Al [c2]

Figure C5
Stratigraphic Section and Geological Map
of Staffa Showing Phillips Et Al Sample Locations

SI

Geographical Sample Point	Mean Column Diameter (m)	Mean Column Side Length t (m)	Column Top Area A (m2)	Mean Internal Angle (radians)
Locality 1	0.775	0.383	0.3157	2.055998
Locality 2	0.834	0.401	0.3709	2.062979
Locality 3	0.918	0.418	0.4329	2.075196
Locality 5	1.135	0.539	0.7520	2.071706
Locality 6	0.746	0.330	0.2836	2.106612
Locality 8	0.878	0.418	0.4326	2.089159
Means μ	0.881	0.414833	0.431283	2.076942

Mean t Multiplier Constant (L^1)	2.123744	1	2.506198	2.094395
Mean t^2 Multiplier Constant (L^2)		1	2.598076	
Normatives (regular hexagon)	0.866025	1	0.504131	2.094395
PSD%(μ,normative)	1.699727	-141.061	-16.891	-0.84034

Table C1
Selected Basaltic Column Statistics for the
Phillips Et Al Extrusion Study

The Indicated Froude Number for Magma Convection at Staffa

Given the assumptions:-

(a)   The "Characteristic Diameter" is a plausible breadth of a magma prism near to the averages determined by Phillips Et Al

(b)   The hexagonal lava prisms were convectional structures of Rayleigh–Bénard type

(c)   The Froude Number is unity at the transition between laminar and turbulent flow

(d)   The Standard Acceleration Due to Gravity is substantially the same as it was in the Tertiary 64-52 million years ago at 9.80665 ms^{-1}

We are in a position to employ Equation C2 to estimate a *maximum* velocity of magma emergence through an active fluid column of magma. This will assist , together with rock fabric analysis, in determining the geological regimes of emplacement.

The Phillips Excircle Diameter Adoption

The Phillips Mean Diameter is 0.881 meters. Hence:-

$$u_{Phillips} = (1).\sqrt{gD_{Phillips}} = \sqrt{9.80665 \times 0.881}$$
$$= 2.93932962595215$$

Equation C8

Therefore the Maximum Uniform Emergence Velocity of the magma is nearly three meters per second when Froude transition equals unity.

The Mean Equal-Perimeter Diameter Adoption

$$t_\mu = 0.414833333$$
Equation C9

$$\rho_P = 0.954929658551372$$
Equation C10

$$D_{EqualPerimeters} = 2.\rho_P.t_\mu = 0.792273306074835$$
Equation C11

The Mean Equal-Perimeter Diameter is 0.792273306074835 meters. Hence:-

$$u_{EqualPerimeters} = (1).\sqrt{gD_{EqualPerimeters}}$$
$$= \sqrt{9.80665 \times 0.792273306074835}$$
$$= 2.78739071839934$$
Equation C12

Therefore the Maximum Uniform Emergence Velocity of the magma is nearly 2.79 meters per second when Froude transition equals unity.

The Reynold's Number

A second key dimensionless parameter of continuum mechanics is the Reynold's Number, Re.
The Reynold's Number is defined by:-

$$Re = \frac{uD}{\nu} = \frac{\rho uD}{\mu}$$
Equation C13

where u is the Uniform Fluid Velocity as before; D is the "Characteristic Length" L; ρ is the Mass Density of the Fluid; ν is Kinematic Viscosity; and μ is Dynamic Viscosity.
As a dimensionlessness check we may write:-

$$Re = \frac{uD}{\nu} = \frac{\rho uD}{\mu} = \frac{ML^{-3}.LT^{-1}.L}{ML^{-1}T^{-1}} = M^0L^0T^0$$
Equation C14

showing that Re is indeed dimensionless. (Or as my very eminent doctoral supervisor, the late DIH Barr insisted upon putting it "non-dimensional").
Having to hand estimates, at least to an order of magnitude, of the emplacement velocity of the Staffa magma, at somewhere near 2.9 meters per second, we can now move forward to estimate the Reynold's Number for the emerging and or convecting

magma knowing that at the Mean Temperature of Basaltic Magma, 1175°C the Density of the Fluid is 2725 kg/m3 and the Dynamic Viscosity is 125 Pascal seconds (kg/(meter×second)):-

$$Re = \frac{uD}{\nu} = \frac{\rho uD}{\mu} = 56.465052$$

Equation C15

Now the Reynold's Number for mobile basaltic magma, largely free of crystals and vesicles at 1175°C is near to 2200 at the laminar-turbulent transition.

57 is very far south of 2200 so the moving basalt magma at Staffa was well within the hydraulic laminar zone, whether it was convecting or emerging unidirectionally.

Item	Symbol	M	L	T	K	Units (SI)	Value
Density of Cold Basalt	ρ_{cold}	1	-3	0	0	kg/m³	2900
Mean Density of Hot Basalt Magma	ρ_{magma}	1	-3	0	0	kg/m³	2725
Temperature of Basaltic Magma (Lower Limit)	T_{lb}	0	0	0	1	°C	1100
Temperature of Basaltic Magma (Upper Limit)	T_{ub}	0	0	0	1	°C	1250
Mean Temperature of Basaltic Magma	T_{am}	0	0	0	1	°C	1175
Dynamic Viscosity of Hot Basalt Magma (Lower Limit) ML1	ν_{lb1}	1	-1	-1	0	Pa.s	110
Dynamic Viscosity of Hot Basalt Magma (Upper Limit) ML1	ν_{ub1}	1	-1	-1	0	Pa.s	140
Mean Dynamic Viscosity of Hot Basalt Magma ML1	ν_{mean1}	1	-1	-1	0	Pa.s	125
Dynamic Viscosity of Hot Basalt Magma (Lower Limit) ML2	ν_{lb2}	1	-1	-1	0	Pa.s	850
Dynamic Viscosity of Hot Basalt Magma (Upper Limit) ML2	ν_{ub2}	1	-1	-1	0	Pa.s	1400
Mean Dynamic Viscosity of Hot Basalt Magma ML2	ν_{mean2}	1	-1	-1	0	Pa.s	1125
Dynamic Viscosity of Water (approximate)	ν_{water}	1	-1	-1	0	Pa.s	1.0016
Linear Expansion Coefficient of Basalt at 100°C	α_{100}	0	0	0	0	N/A	0.000001
Linear Expansion Coefficient of Basalt at 1100°C	α_{1100}	0	0	0	0	N/A	0.000017
Mean Terrestrial (Standard) Acceleration Due to Gravity	g	0	1	-2	0	m/s²	9.80665
Maximum Thickness of Staffa (Fingle's Cave) Colonnade Basalt	H_{max}	0	1	0	0	m	40
Thickness of Staffa (Phillips et al Unit 2) Colonnade Baslt	H_{unit2}	0	1	0	0	m	8
Minimum Thickness of Staffa (Phillips et al) Colonnade Basalt	H_{min}	0	1	0	0	m	1
Subaerial Velocity of Mafic Lava Flow (0.4 km/h)	v_{mafic}	0	1	-1	0	m/s	0.277778
Froude Number	Fr	0	0	0	0	N/A	
Reynold's Number	Re	0	0	0	0	N/A	

Please add 273.15 to convert Degrees Celsius to Degrees Kelvin
ML1: Mauna Loa Magma (5% crystals: 0% vesicles) (Harris and Allen)
ML2: Mauna Loa Magma (27.5% crystals: 50% vesicles) (Harris and Allen)

Table C2
Gross Physical Data for Columnar Basalt and Basaltic Melts at Staffa and Mauna Loa

INDEX

Definite, 56, 175, 178

degree, 16, 28, 65, 75, 104, 111, 118, 165, 175, 178

Degree, 27, 28, 29, 121, 175

dendroids, 146

denominator, 22, 115, 118

Density, 150, 155, 197, 198

Dependent, 27

Determination, 66, 76

DFAC, 169, 174

DFACEVEN, 170

DFACODD, 170

diameter, 21, 189, 190

Diameter, 14, 36, 173, 190, 191, 193, 194, 196, 197

Differential, 22, 27, 28, 29, 32, 55, 150

Digit-Extraction, 64

digits, 14, 21, 64, 65, 73, 76, 111, 112, 132, 134, 171, 172

DIH Barr, 197

dimensionless, 52, 149, 188, 197

Diophantine, 111, 115

Division, 112

Doris, 15

Double Factorial, 169, 174

Double-Traverse, 154

drag, 187, 190

Dynamic, 187, 197, 198

Edition, 2, 3, 43, 158, 163

Egyptian, 61, 71, 75, 76

Eight, 163

Einstein, 31

element, 37, 105, 122

Emergence, 196, 197

empirical, 27

Empirical, 30

Eneolithic, 20

energy, 149, 187, 188

engine, 21

engineering, 19, 105

Englishmen, 18

Enlightenment, 187

Epigraph, 7, 157

equal, 19, 22, 29, 122, 154, 190, 191

equation, 24, 29, 64, 75, 82, 90, 104, 118, 175, 179

Equation, 14, 20, 22, 23, 27, 28, 29, 30, 31, 32, 36, 37, 38, 39, 40, 41, 42, 43, 44, 45, 47, 48, 49, 50, 51, 52, 53, 55, 56, 57, 58, 59, 60, 61, 62, 63, 64, 65, 69, 70, 71, 72, 73, 74, 75, 76, 79, 80, 81, 82, 83, 87, 88, 89, 90, 91, 95, 96, 97, 98, 99, 100, 101, 102, 103, 104, 105, 106, 107, 108, 109, 110, 111, 112, 113, 114, 115, 118, 119, 120, 121, 122, 123, 124, 125, 126, 127, 128, 129, 131, 132, 133, 134, 135, 136, 137, 138, 139, 140, 141, 142, 143, 146, 150, 155, 169, 170, 171, 172, 173, 175, 176, 177, 178, 179, 180, 181, 182, 183, 184, 189, 193, 194, 196, 197, 198

error, 11, 32, 123, 129, 141, 142, 173, 194

escribed, 190

Estimate, 23, 63, 83, 89, 134, 137

Estimation, 194

Euclid, 20

Euclidian, 150

Euler, 23, 30, 31, 32, 33, 62, 64, 65, 66, 74, 79, 80, 89, 132, 135, 136, 138, 141, 154, 160, 161

Euler Circuit, 154

Europe, 23, 69

Excircle, 190, 196

exoskeleton, 15

Expansion, 23, 143, 173

explosive, 79

Exscribed, 190

Extended, 70, 75, 76, 159, 162, 177, 178, 179, 182, 183

Extreme, 20

Extrusion, 195

fabric, 189, 196

facies, 17, 148, 150

FACTDOUBLE, 169, 170, 174

Factorial, 169, 173

faith, 11, 13

Famous, 37, 155, 187

faults, 150

feeding, 15

fido, 62, 63, 64, 141

Fiducial, 131, 141, 169
File, 2, 157, 168
First, 2, 3, 22, 27, 28, 29, 32, 55, 58,
    66, 88, 100, 137
fitful, 13
fitted, 29, 31, 55, 178
Five, 32, 132, 159
flight, 187
floor, 13, 15
flow, 168, 187, 188, 189, 196
Flow, 188
fluid, 13, 149, 151, 187, 188, 190,
    196
fluids, 187
fodinichnial, 151
force, 187
forces, 19, 79, 189
Form, 137, 178
Formation, 165
Formula, 23, 40, 63, 65, 74, 75, 76,
    100, 113, 137, 138, 178
Formulae, 132, 177, 179
Formulas, 159, 162, 163, 164
fossil, 15, 18, 148, 157
Four, 4, 37, 88, 155, 159
fractal, 16
Fraction, 95, 113, 155
Fractions, 71, 75, 76, 96, 99, 112,
    146, 163
Free, 5, 157, 159, 160, 161, 162,
    163, 164, 166, 167, 168
fresh, 18
friction, 190
Froude, 149, 150, 166, 168, 188,
    189, 196, 197
FUN18, 181
function, 21, 22, 27, 29, 52, 55, 133,
    136, 137, 146, 162, 169, 170
Function, 22, 29, 57, 87, 88, 133,
    162, 169, 173, 175, 177
functions, 22, 43, 69
Functions, 22, 69, 159, 162
Gaelic, 192
Gathering, 121
Gaze, 11
gender, 18
Gendering, 18

geological, 150, 151, 189, 196
Geological, 167, 195
geology, 149, 167
geometric, 19, 190
Geometry, 21
ghost, 148
Glory, 5
glyph, 18
God, 5, 11, 18, 21, 145
Goldbach, 79
Golden, 20
goniometric, 146
goniometry, 149, 194
gorilla, 145
Grade, 27
gradient, 151
Gradient, 55, 66, 76
graph theory, 154
graphically, 27
gravitational, 189
Gravity, 150, 189, 196
Greece, 21
Greek, 15, 20, 61
Greeks, 20, 61
grooves, 15
group, 114
groups, 14
Guillera, 83
half, 36, 65, 90
Handbook, 159, 162, 163
hardware, 149, 171
Harlan J Brothers, 80, 161
Haunted, 20
Hazards, 112
Hebrides, 192
height, 187, 190, 192
Height, 150, 189
Helminthoida, 15, 157
heptagon, 189
hermaphrodite, 18
heteromorphism, 151
hexadecimal, 132
hexagon, 148, 150, 155, 189, 190,
    193, 194
Highland, 192
Hilbert, 16
Hippasus, 21

Packing Density, 155
Packing Fraction, 155
Page, 7, 8, 9
Paleodictyon, 148, 149, 150, 151, 152, 154, 155, 166
Paleodictyon nodosum, 148
Paleozoic, 15, 17
Papyrus, 61
parabolic, 177, 190
Paradox, 19
parameters, 65, 76, 177, 178, 188, 193
Partial, 112, 163
Passive, 19
pattern, 14, 15, 16, 149, 178
pelagic, 13
pentagon, 36, 189
pentagonal, 41, 151
Pentagram, 20, 35, 40
Percentage, 58, 59, 70, 169, 193
Periclean, 20, 21
Perimeter, 14, 190, 196, 197
Permian, 148
Peter Borwein, 132
Pf(), 133, 135, 136, 137, 140
phantasm, 13
Phenomenal, 19
phenotype, 18
Phi, 4, 11, 20, 158
Phidias, 11, 20, 21, 25, 36, 51, 131, 139
Phillips Et Al, 192, 193, 195, 196
phyla, 13, 15
phylum, 18, 154
physics, 155, 188
Pi, 4, 11, 20, 62, 64, 113, 116, 132, 133, 134, 140, 158
Pi and Phi, 4, 11, 20, 158
Pippenger, 87, 91, 162
plane, 15, 149, 150
planes, 15, 150
Plouffe, 64, 99, 111, 118, 132, 142, 163
plug, 22, 122
plus, 29, 55, 139
point, 14, 15, 23, 27, 29, 30, 44, 58, 59, 66, 81, 114, 154, 177

Point, 29, 32, 55
polychaete, 18
polygon, 61, 155, 189, 194
Polygonal, 192
Polynomial, 27, 28, 56, 101, 102, 112, 175, 178
position, 27, 115, 196
positive, 30, 31, 55, 175, 178
Potent, 2
potential, 149, 188
Power, 81
Powering, 81
Prefix, 2
pressure, 13, 148
prime, 27
primordial, 16
prism, 151, 196
Procedural, 112
procedure, 30
Productions, 2, 3, 12, 158
program, 55
programming, 171
programs, 29
Province, 153, 192
PseudoBBP, 138
Psi, 75
Publication, 7
Putterer, 146, 165
Pythagoras, 21, 25, 37, 69, 128, 131, 141, 160, 164
quadratic, 32, 56, 75, 82, 91, 175, 176, 177, 178, 179, 180
Quadratic, 28, 69, 175, 177, 182, 183, 184
QuadTerm, 99
Quartic, 177
Queen, 19
Quintic, 177
radius, 21
Radius, 36, 39, 40, 194
radix, 22, 23, 30
Radix, 22
Ramanujan, 62, 63, 65, 66
Ramanujan-Sato, 62, 63, 65, 66
Raphson, 22, 23, 24, 30, 31, 32
Rate, 29, 55, 56, 65, 66, 75, 76

Split, 99

square, 12, 20, 21, 22, 28, 30, 31, 52, 138, 155, 171

Square, 11, 20, 21, 22, 23, 26, 33, 36, 37, 52, 69, 128, 131, 132, 137, 160, 164

Staffa, 153, 167, 168, 192, 194, 195, 196, 197, 198, 199

Statistics, 195

Stegun, 159, 162, 179

Stone, 15, 18, 23

straight, 21, 28, 31, 57, 65, 75

strategy, 15, 154

Stratigraphic, 195

stress, 149

sub-parabolic, 190

Successor, 23

sum, 27, 89, 127

Sumerians, 20

summation, 30, 71, 105, 177, 181

summer, 18

surd, 35

surface, 14, 18, 149, 150, 151, 187, 188, 190

Swiss, 30, 62

symbolic, 43, 113

Syracuse, 61

Table, 7, 24, 26, 30, 31, 32, 65, 66, 68, 76, 78, 82, 86, 88, 91, 93, 114, 115, 116, 169, 174, 179, 180, 182, 183, 184, 193, 195, 199

Target, 141

Taylor, 23, 30, 32, 33, 73, 74, 75, 76, 142, 143

Taylor Series, 23, 30, 33, 73, 74, 75, 76, 142, 143

Temperature, 198

Ten, 165

tensional, 150

Term, 29, 55, 56, 70, 117, 175

terrestrial, 18

Tertiary, 153, 192, 196

textual, 19, 21

Theodorus, 12, 131

Theorem, 23, 30, 37

theoretical, 61, 172, 173

thermodynamic, 149

thickness, 187

Thomas, 13, 157

Three, 4, 16, 26, 97, 98, 131, 158

time, 15, 23, 27, 43, 66, 133, 145, 151, 155, 171, 189

Title, 7

tolerance, 22, 169

traces, 15

tracks, 15, 18

Transcendent, 177

Transcendental, 28

Transform, 74

Trapezoidal, 57, 58

traps, 19

Trawls, 19

trendline, 64, 75

triangle, 47

triangular, 151

trigonometry, 21, 35

triode, 171

triple, 95

triple-term, 95

Triterm, 97, 98

Turbulent, 188

tutorial, 122

Tutorial, 2, 3, 12, 158

Two, 4, 11, 21, 26, 37, 57, 69, 98, 112, 128, 131, 158

Uniform, 196, 197

uniform velocity, 190

unitary, 21

unity, 21, 29, 36, 52, 59, 112, 113, 115, 121, 139, 189, 190, 193, 196, 197

univariate, 113

unknown, 13

Upper, 56, 175, 178

value, 14, 21, 29, 30, 36, 39, 41, 43, 48, 50, 55, 59, 61, 62, 69, 82, 99, 105, 114, 115, 126, 137, 169, 170, 176

Value, 29, 30, 55, 64, 133, 141

values, 24, 31, 64, 88, 111, 115, 118, 121, 122, 131, 142, 179, 185, 189

Values, 55, 88, 131

variable, 175

Variable, 27, 28

www.ingramcontent.com/pod-product-compliance
Lightning Source LLC
Chambersburg PA
CBHW040139200326
41458CB00025B/6315